Industry 4.0 - Current Status and Future Trends

Edited by Jesús Hamilton Ortiz

Published in London, United Kingdom

IntechOpen

Supporting open minds since 2005

Industry 4.0 - Current Status and Future Trends
http://dx.doi.org/10.5772/intechopen.86000
Edited by Jesús Hamilton Ortiz

Contributors
Felicita Chromjakova, Ercan Oztemel, Samet Gursev, Wahyu Caesarendra, Vigneashwara Pandiyan, Georg Spoettl, Jesus Hamilton Ortiz, Leonardo Zambrano, Núbia Gabriela Pereira Carvalho, Edson Cazarini, Ana Paula Pereira Carvalho, Ana Claudia Pereira Carvalho

First published in London, United Kingdom, 2020 by IntechOpen
IntechOpen is the global imprint of INTECHOPEN LIMITED, registered in England and Wales, registration number: 11086078, 7th floor, 10 Lower Thames Street, London, EC3R 6AF, United Kingdom
Printed in Croatia

British Library Cataloguing-in-Publication Data
A catalogue record for this book is available from the British Library

Additional hard and PDF copies can be obtained from orders@intechopen.com

Industry 4.0 - Current Status and Future Trends
Edited by Jesús Hamilton Ortiz
p. cm.
Print ISBN 978-1-83880-093-2
Online ISBN 978-1-83880-094-9
eBook (PDF) ISBN 978-1-83880-086-4

An electronic version of this book is freely available, thanks to the support of libraries working with Knowledge Unlatched. KU is a collaborative initiative designed to make high quality books Open Access for the public good. More information about the initiative and links to the Open Access version can be found at www.knowledgeunlatched.org

For EU product safety concerns:
IN TECH d.o.o., Prolaz Marije Krucifikse Kozulić 3, 51000 Rijeka, Croatia, info@intechopen.com or visit our website at intechopen.com.

Meet the editor

Dr. Jesús Hamilton Ortiz is a professional with wide experience as a professor and researcher in computer engineering telecommunications and mathematics. He is an international reviewer of recognized journals (IEEE, Elsevier, IAJIT, etc.), and an expert editor in: ad hoc networks, mobile networks, computer technology, telecommunication networks, wearables, Industry 4.0, drone swarms, and algorithms. With more than 300.000 downloads, he has published more than 100 articles and 7 books. He has also been a thesis advisor at the undergraduate and postgraduate level in: telematics, computer, telecommunication, electronic engineering programs, and is currently a professor at the UNAD University and CEO in Closemobile R&D.

Contents

Preface

The fourth industrial revolution happened as an evolution process in terms of technological advances, and it represents the social, economic, and technological changes that affected all developed countries. The fourth and last industrial revolution brought advice and forecast into the system. This new age of revolution is based on the creation of an integrative and collaborative environment, with the intention to add cyber physical systems, and to consider the customer as part of the production process. In this fourth industrial revolution, the analysis and testing of the product is done before it goes to the market, not after, which allows for the opportunity to predict what is going to happen before it happens, improving the management of the investment that is made in the production.

The main outcome of the fourth industrial revolution was the set-up of the Intel industries, where it is not just about data compilation, but keeping in mind the importance of the interaction between all this information, being allowed to connect it with different information sources, and being able to create a controlled system with a high-speed and flexible type of production.

The fourth industrial revolution is the last advancement of humanity. It started with technological advances that the population have in their hands nowadays. It is mainly based on the digitalization, automation, personalization, customization, and integration between IT management systems and production systems of the factories through the use of the internet of things (IoT). It is also paired-up with cyber-physical systems. These systems (or pillars) will be the autonomous kind of systems, with the capacity to make their own decisions and apply machine learning. These systems will also collect data in real-time that will be analyzed and saved into the cloud. Industry 4.0 will allow us to exploit pillars such as the internet of things (IoT), Big Data and data analytics, augmented reality (a virtual representation of the real world), cyber security, collaborative robots, additive manufacturing, cloud computing, artificial intelligence, and finally, 5G networks.

In addition, we can say that industry 4.0 could be capable of decreasing the production costs, logistic costs, and quality management costs. The industry products have implemented sensors (IoT) connected to the network, capturing and storing information constantly into the cloud (or a server), for the purpose of being processed later. This means that the information is closely related to the pillars of Big Data and data analytics. Also, the IoT is related to the cyber security pillar because of the interaction between the devices connected to the internet, communicating between themselves constantly, making them highly vulnerable to a computer attack.

Also, investors in Industry 4.0 will change technological aspects, product development processes, marketing, logistics, manufacturing, after-sales services, and security. At the beginning of the fourth industrial revolution, all economic, financial, social, cultural, and environmental models received a great impact; they must now adapt to this new revolution. One of the biggest challenges is the negative impact on the environment of the traditional production model; it affects human life but there is not enough awareness to counter it. Perhaps people think that it has affected others and not themselves.

Pollution includes the large volumes of plastics produced that go to the ocean; food that is thrown into the trash; computers, telephones, cars due to overproduction exceeding the demand; excessive electricity consumption; excessive use of gas vehicles; poor policies to mitigate and control environmental pollution; excessive CO_2 emissions including computer equipment and computers. For all these reasons, Industry 4.0 is an opportunity to modify all these superficially considered aspects of the traditional economic model. There must be awareness that natural resources are limited and it cannot be allowed to continue destroying them as is being done so far. The new economic model must consider that the natural resources must be protected and the environmental impact must be minimized. It should unify these two concepts: create wealth and wellness but simultaneously cause minimal environmental impact.

Similarly, one of the big bets for Industry 4.0 is the productivity increase generating greater wealth, but a consequence of that wealth must be greater equity in the resource distribution as a result of good management of the fiscal and tax control by the ministry. A serious mistake would be to focus on the overproduction and natural resources exploitation, and neglect the impact it has on nature. Some economists call it circular economy, biophysical economics, or ecological economics. This economy is based on several principles: preserve natural capital by controlling natural and renewable resources, optimize the resource use, promote a more efficient system that eliminates external and internal aspects in the productive chain and particularly that has minimum environmental impact, design no waste products, elastic and flexible diversity, renewable energy increase, consider everything as a system, and re-evaluate the costs and prices management.

In conclusion, the vision of current and future industry and society with intelligent industries, autonomous cars, intelligent cities, etc. is integrated. The actions against climate change are: minimize all negative issues in business world and management systems. A more supportive society, with greater cooperation between people, communities and countries, greater trust among people, minimum corruption in public management, a more equitable resource distribution is an ideal thought; regarding markets, the basis of the production process would be greater trust and credibility between clients and companies.

This book shows a vision of the present and future of Industry 4.0 and identifies and examines the most pressing research issue in Industry 4.0. Containing the contributions of leading researchers and academics, this book includes recent publications in key areas of interest for example; a review on the Industry 4.0: What is the Industry 4.0, the pillars of Industry 4.0, current and future trends, technologies, taxonomy and some case studies (A.U.T.O 4.0, stabilization of digitized process). Also, this book provides an essential tool in the process of migration to Industry 4.0. This book is suitable as a text for graduate students and professionals in the industrial sector and general engineering areas.

The book is organized into two sections:

1. **Reviews:** Industry 4.0, what is it, Industry 4.0: current status and future trends, Industry 4.0 what is your potential, a taxonomy of industry and related technologies

2. **Case Studies:** A.U.T.O 4.0, Stabilization of digitized processes, framework for integrating Brownfield Systems to establish intelligent manufacturing architecture

Industry 4.0 is likely to play an important role in the future society. This book is a good reference on the latest Industry 4.0 information, including some case studies. Each chapter is written by expert researchers in the sector, the topics are broad and show the concept or definition of industry 4.0 to a future society 5.0.

Jesús Hamilton Ortiz, PhD
CEO,
Closemobile R&D,
Madrid, Spain

Full Professor,
UNAD,
Bogotá, Colombia

Section 1

Review

Chapter 1

Industry 4.0 - What Is It?

Núbia Gabriela Pereira Carvalho and Edson Walmir Cazarini

Abstract

The industry 4.0 is a new industrial model that characterizes the Fourth Industrial Revolution. This advanced manufacturing model is represented by intelligent, virtual, and digital performance in large-scale industries and emerges as a disruption to the three industrial revolutions that occurred before it. The new industrial model itself includes a factory-wide integrated structure and potential technologies in various areas of industry activity, and these technologies are intrinsic to industry 4.0 design principles, which are also responsible for ensuring the innovative performance of this new industry. With this explanatory context, the objective of this paper is to present what is really industry 4.0, its origin, as well as its main characteristics. In this way, knowing what the Fourth Industrial Revolution is, the readers will be able to better understand the content presented in the next chapters about the diverse potentials of the new industry.

Keywords: industry 4.0, Fourth Industrial Revolution, industrial model 4.0, advanced manufacturing, design principles

1. Introduction

The history of the industry since its inception is marked by the acquisition of great knowledge, events, and discoveries that have changed the structure of cities, their populations, the types of products they offer and the way they offer them, innovation in the development of production processes, and the way employees work, among many other aspects.

The industry at the time of its emergence can be verified as a result of the combination of three fundamental aspects: knowledge, experimentation, and entrepreneurial innovation. These aspects have effectively contributed as a potential medium for offering jobs and developing new products to people and their needs.

Thus, the industry has its evolution distributed in different stages and of equal relevance. Such stages are represented by the First, Second, Third, and, imminently, Fourth Industrial Revolution. It would be misleading to say that industry 4.0 is a new model that is totally unaware of the industrial revolutions that occurred earlier. In fact, industry 4.0 fully utilizes all the foundations of previous industrial revolutions, however, with higher rates of integration, digitization, virtualization, technologies, and rapid response times to stimuli.

Just as the industry has undergone many changes and improvements in each of its stages, the products developed, the ways of distribution, the customer demands, the work of people, and the internal and external aspects have changed. Knowing the industry 4.0 and all its capabilities in different areas is essential for the full use of its potential, and also for those looking to achieve a form of collaboration in the new industry, it is important to know all the new opportunities it presents.

In this context, the purpose of this chapter is to introduce industry 4.0 and its main features. For this, this chapter is structured in three next sections, among which the first refers to the historical evolution of the industry, while the second refers to the origin of industry 4.0 and its main understandings, and finally, the third refers to the design principles of the new industrial model.

2. Historical evolution of industry

Considering the combination of the three fundamental aspects that contributed to the emergence of industry, namely, knowledge, experimentation, and entrepreneurial innovation, industrial revolutions could have their consolidation designed. The first aspect, knowledge, refers to the range of understanding of scholars and inventors of decades ago, who researched, investigated, sought possibilities for great new discoveries at the time, and was able to increase their ability to understand and assimilate. The second aspect, experimentation, is the process of making explicit the tacit knowledge of scholars and collaborators, in order to prove and prove possibilities and unfeasibilities. The third aspect, entrepreneurial innovation, is the confidence of employers of the time to believe in and invest in new discoveries, work models, and ways of producing and expanding their business on scales never before imagined.

The industrial revolutions cannot be reported only by inventions or discoveries of energy sources, new machines, materials, or methods. However, these factors are of utmost importance for the development of the economy in the last two and a half centuries. Decades ago machines such as hydraulic mills already existed. However, the spread of these machines, called machinery, characterizes and distinguishes this period from previous periods [1]. **Figure 1** [2] presents the three industrial revolutions that have already taken place with their respective highlights.

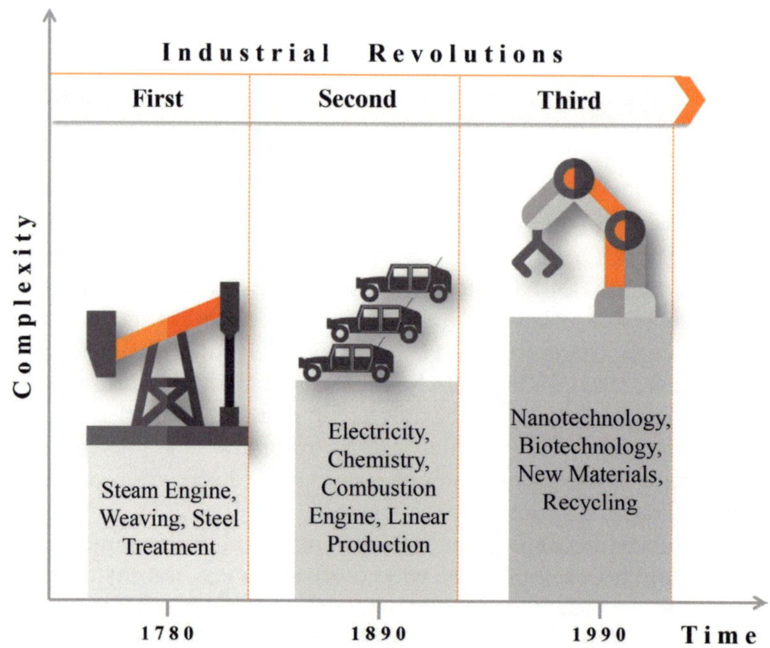

Figure 1.
Industrial revolutions and their highlights. Source: Adapted from [2].

Until the First Industrial Revolution, from the mid-eighteenth century, the sources used were water, animal energy, and wood, which was the main building and combustion material. In construction, iron became one of the most prevalent materials, while coal was the first fuel, and steam one of the main engines, all interacting with each other forming a new technical system at the time [3].

The development of industrial society was observed at three distinct times. The first of these corresponds to the so-called First Industrial Revolution, which was driven by the creation of the steam engine and had as its greatest theorist Adam Smith, who was also the creator of economic liberalism [4].

The First Industrial Revolution lasted until the last decades of the nineteenth century, and during this period significant changes occurred. The factory production has been expanded to countries other than England as well as to other consumer goods segments. The mid-nineteenth century saw the revolution in transport and communications, including the spread of the railways, the telegraph, and transoceanic steamboat navigation in steel hulls. Activities that went beyond borders, which were considered a mixture of enterprise and adventure, are made possible and commercially and financially integrated into the logic of business expansion [5].

After the First Industrial Revolution, the so-called Second Industrial Revolution began in the second half of the nineteenth century, which included a series of developments in the chemical, electrical, oil, and steel industries and also encompassed other key developments during this period like steam-powered steel ships; airplane development; canning of food; mass production of consumer goods; mechanical refrigeration, as well as other preservation techniques; and the creation of the electromagnetic telephone. This era also marks the advent in Germany and the United States together with France and the United Kingdom as potentialities in the industrial environment. In the United States, the Second Industrial Revolution is commonly associated with the scientific management proposal studied and applied by Frederick Winslow Taylor [6].

The transition to the Second Industrial Revolution is consolidated with the inclusion of new sources of raw materials and energy, especially electricity, and increasingly oil. It is at the end of the nineteenth century that a certain cyclical pulse can be seen in the world economy. In the early decades of the twentieth century, the revolution in communications and transport intensified with the spread of the automotive and aeronautics industries, as well as telephony and radio transmissions [5].

In accordance with the statement of the previous authors, the author [1] considers that from the last half of the nineteenth century, it can be affirmed that there was a Second Industrial Revolution. While the former was based on iron and steam power from coal, the latter was based on electricity and steel, providing relevant developments in communications, chemistry, and the use of petroleum. But these innovations, initially and generally, did not fully replace the old ones, so they only began to stand out, as their full realization took place only in the twentieth century.

The Third Industrial Revolution started in the 1970s [7] and characterizes the structuring of capitalism, so the economic system presented cooperation as a new paradigm, arguing that the objectives of an organization can only be achieved together. Given this, there was a high hiring of companies; the development of these happened more and more driven to get strong joints with the same proportion of reciprocity. This third revolution presented a different way of managing the economy, highlighting a more horizontal, more flexible negotiation, where companies come together to achieve more precise and consolidated economic action [8].

The Third Industrial Revolution developed the role of the instrumentation of the financial economy, called the market economy, providing accelerated development in the media and transport, as well as allowing the global integration consisting of globalization [6]. For [7], this revolution was well known for the technologies that were employed in Japan in the mid-1970s, including biotechnology, computer science, Computer Numerical Control (CNC), microelectronics, and telematics integrated system, among others.

3. Industry 4.0: origin and key understandings

The term "Industry 4.0" comes in the context of a new industrial revolution, which emphasizes and includes the latest technological innovations and aids in both fast and customized production. The term originated in Germany in 2011, referring to changes directly linked to automation fields integrated with information technology [9].

With the aim of promoting manufacturing automation and, consequently, increasing productivity through "smart factories," the still imminent Fourth Industrial Revolution can be characterized by the integration between the Internet and production processes, with the aid of smaller sensors and artificial intelligence applied to machines [10].

There are numerous and distinct definitions and nomenclatures for industry 4.0; however, the definition set out in this chapter is that industry 4.0 is an advanced manufacturing model that includes within itself an extensive set of technologies not necessarily unpublished but integrated with each other and with the whole industry which is characterized by its high virtual, digital, and technological performance. The industry 4.0 is still a new industrial model with disruptive characteristics in relation to environmental performance as it is considered a sustainable manufacturing model, as well as human work by offering a set of potential technologies that help the work performed by people in the industry, eliminating aspects such as intense physical effort, and market positioning, as industry 4.0 has an excellent response time to internal and external stimuli.

By 2014 in Germany, industries were already assessing their readiness for industry 4.0 implementation. At that time at least 41% of German companies were aware of the industry 4.0 term and so have started some concrete initiatives. But the road to go is long and some industries still do not know the term. This applies in particular to small-scale industries, where 44% did not know what industry 4.0 is. On the other hand, the new industrial model is well known in larger companies, where only 17% did not know the term. There is also a delay in industry 4.0 implementation plans between large industries and small and medium enterprises (SMEs). Almost 20% of original equipment manufacturers had solid implementation strategies; while there is a huge volume of SMEs in Germany, only 17% were equipped with implementation strategies [11].

According to [12], industry 4.0 is not only a technical challenge but also a reality that will significantly change the organizational structure of companies. Not intended to provide a comprehensive listing, the authors exemplify five visions for disruptive industry 4.0 change:

- New level of socio-technical interaction: Autonomous and self-organized production resources carry out planning processes in value chains between organizations.

- Smart products: The products and the tolerable operating parameter of a certain production process are mutually known. These products can be grouped to optimize production.

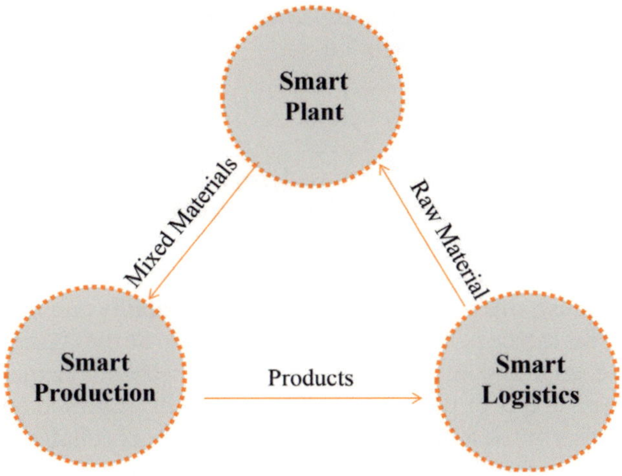

Figure 2.
Key aspects of industry 4.0. Source: Adapted from [13].

- Individualized production: Flexible reconfiguration enables industries to consider the specific characteristics of customer demand and product during design, planning, production, and recycling phase.

- Autonomous control: Employees control and configure intelligent production resources based on targets sensitive to the present context.

- Product design controls product-related data: Product-related data becomes a central feature in managing its product life cycle.

According to [13], industry 4.0 is a junction of three general aspects to consider: smart plant, smart production, and smart logistics. According to [14] the "smart plant" has evolved from the digital factory and is a key component of the smart infrastructure for the future, with an emphasis on highly potential production systems and processes, as well as performing network distribution in production facilities, which constitutes the factory layout. **Figure 2** [13] presents the integration between the three aspects mentioned above.

The "intelligent production," primarily related to industry-wide production logistics management, coupled with human-to-machine (H2M) interaction as well as the application of 3D/4D technology in industrial processes, is a high-capacity industrial chain with a high capacity to be flexible, personalized, and actively maintained in a network. The "intelligent logistics," especially through the Internet and logistics resource integration networks, cooperates with the high performance of logistics resource efficiency on both the supply and demand side, so it is can obtain a service match for logistical support. These three aspects are independent of each other, which with coordination and mutual cooperation constitute the industry 4.0 production system. It is noteworthy that smart manufacturing is the core of this new industry, whose purpose is to create customized products for customers [15].

4. Industry 4.0 design principles

According to [16], the industry 4.0 encompasses six design principles in its framework, which are called decentralization, virtualization, interoperability, modularity, real-time capability, and service orientation. These principles are called "design principles" because they contribute to the design or transition process of a common industry, or 3.0, to industry 4.0.

4.1 Decentralization

The first principle, decentralization, is understood in industry 4.0 as the greater ability of local companies and specific operations, as well as those carried out by machines, to make their own decisions on their own. Rather than using central computers or passing a decision hierarchically, enabling and allowing local operators to respond to changes and readjust, this principle provides more flexibility and makes it easier to use expertise. This can be perceived as a decomposition of the classical production hierarchy for the change of companies as decentralized self-organizations [17].

However, the principle of decentralization cannot be observed only for machines, as it refers to the autonomy granted to people as collaborators in industry 4.0. They have greater freedom to identify aspects, analyze parameters, and carry out decision-making whenever necessary, aiming at the common good for their area of activity in the industry as well as for its fullness.

4.2 Virtualization

For [18], the principle of virtualization is that by using machine-to-machine (M2M) monitoring and communication, a virtual twin can be abstracted from the industry. The sensor data is linked to virtual plant models and simulation models. Thus a virtual copy of the physical world can be created. In case of failure, an employee can be notified. In addition, all necessary information, such as next work steps or security provisions, remains available.

The virtualization in industry 4.0 is used by people as highly potential tools to aid human work. This principle streamlines the time, analysis, and decision-making of employees and established teams by providing, sharing, and synthesizing information virtually, quickly, and in real time.

4.3 Interoperability

The principle of interoperability in the industry 4.0 manufacturing environment is that cyber-physical system (CPS) comprises intelligent machines and intelligent storage systems and facilities capable of autonomously exchanging information, initiating actions, and controlling each other independently. The embedded manufacturing systems are vertically connected with business processes internal to industries, and horizontally, with the value chain, by connecting software and programs [19].

The interoperability is also linked to the work performed with H2M interaction, consisting also in the ability of people to work harmoniously with the machines, so that the committed effort is realized in sync in all industrial activities.

4.4 Modularity

The principle of modularity involves modular systems that can flexibly adapt to changing requirements by replacing or expanding individual production modules, making adding or removing modules much easier. These modular systems can thus simply be adjusted in case of seasonal fluctuations or changes in product production needs, as in the case of including new technologies [20].

Thus, production can always adjust to environmental, systemic, and changing customer demands without error, lost productivity, or customer dissatisfaction.

4.5 Real-time capability

To define the principle of real-time capability, [19] states that in the manufacturing process, intelligent machines with specific software will automatically adapt to the process and decision-making by CPS to the productive needs, thus monitoring the product quality in order to make decisions at every moment of need. This interconnection will minimize misuse of resources, waste, material waste, and increase energy efficiency.

The real-time capability principle is one of the most outstanding aspects of industry 4.0 as it is responsible for ensuring that the industry has the best possible response time to internal and external stimuli by sharing, receiving, and analyzing data and information in real time.

4.6 Service orientation

The principle of service orientation, according to [20], is characterized by the availability, through the Internet, of human, business services, and CPS, which can be used by other stakeholders, facilitating the creation of product-service systems (PSS), also known as product-service. They can be offered internally to and outside the organization.

In this way, industry 4.0 preserves its network performance in partnerships with all its stakeholders, whether customers, partner industries, and suppliers, among many others. Everyone can have access to useful services, products, and information about the industry using virtual and digital platforms available at all times.

5. Conclusion

This chapter presents the context of the historical evolution of industry 4.0, its origin, its main characteristics, and finally the design principles that are intrinsic to the new industrial model. Industry 4.0 is an imminent model of advanced manufacturing that has not yet massively consolidated itself across all industries in many countries. It requires reflection on impacts, analysis of possibilities, and unfeasibilities and practical studies with academic and business partnerships. The potential of the new industry is numerous for all its areas and different segments; its complete transition or implementation requires knowledge and good use.

Author details

Núbia Gabriela Pereira Carvalho* and Edson Walmir Cazarini
University of São Paulo (USP), São Carlos, Brazil

*Address all correspondence to: nubia.carvalho@usp.br

References

[1] Dathein R. Inovação e Revoluções Industriais: uma apresentação das mudanças tecnológicas determinantes nos séculos XVIII e XIX. Decon-Textos Didáticos/UFRGS. 2003;**2**:1-8

[2] Dombrowski U, Wagner T. Mental strain as field of action in the 4th industrial revolution. Procedia CIRP. 2014;**17**:100-105. DOI: 10.1016/j.procir.2014.01.077

[3] Gille B. Histoire des Techniques. Paris: Pléiade, Gallimard Editions; 1978. 1652 p

[4] De Masi D. A sociedade pós-industrial. 3rd ed. SENAC São Paulo: São Paulo; 1999. 444 p

[5] Guedes C, Rosário J. Informação e Conhecimento: os impactos na reorganização do mercado e do trabalho. Desenvolvimento em Questão. 2002;**3**:9-34. DOI: 10.21527/2237-6453.2005.5.9-34

[6] Contreiras PA. Quarta Revolução Industrial: Um estudo de caso realizado na empresa Lix de Tecnologia. Revista Gestão. Inovação e Negócios. 2015;**1**:79-97. DOI: 10.29246/2358-9868.2015v1i1

[7] Costa A, Silva F, Veloso M, Junior R, Ferreira W, Feitosa J. A influência da indústria na economia do município de Rolim de Moura - RO. Revista Farol. 2017;**5**:147-161

[8] Fidelis C, Reis M. A desconsideração da personalidade jurídica à luz do novo código de processo civil. Direito, Cultura e Cidadania. 2016;**1**:1-26. DOI: 10.26547/2236-3734.dcc.v6i1.71

[9] Kagermann H, Anderl R, Gausemeier J, Schuh G, Wahlster W. Industrie 4.0 in a Global Context: Strategies for Cooperating with International Partners (Acatech Study) [Internet]. 2016. Available from: https://www.acatech.de/wp-content/uploads/2018/03/acatech_STU_engl_KF_Industry40_Global_01.pdf [Accessed: 29 August 2019]

[10] Schwab K. The Fourth Industrial Revolution. Genebra: Currency; 2016. 192 p

[11] Weiss M, Zilch A, Schmeiler F. Industrie 4.0 status Quo und Entwicklungen in Deutschland. Eine Analyse der Experton Group; 2014

[12] Kagermann H, Wahlster W, Helbig J. Recommendations for Implementing the Strategic Initiative Industrie 4.0. Germany: ACATECH National Academy of Science and Engineering; 2013. 97 p

[13] Zhou K, Liu T, Liang L. From cyber-physical systems to industry 4.0: Make future manufacturing become possible. International Journal of Manufacturing Research. 2016;**11**:167-188. DOI: 10.1504/IJMR.2016.078251

[14] Kagermann H. Change through digitization - value creation in the age of industry 4.0. Management of Permanent Change. Springer. 2015:23-45. DOI: 10.1007/978-3-658-05014-6_2

[15] Anderl I. Industrie 4.0 - advanced engineering of smart products and smart production, technological innovations in the product development. In: 19th International Seminar on High Technology; 09 October 2014; Piracicaba. Brazil; 2014. pp. 1-14

[16] Hermann M, Pentek T, Otto B. Design Principles for Industrie 4.0 Scenarios: A Literature Review. Working Paper, No. 01-2015. Dortmund, Germany: Technical University of Dortmund; 2015

[17] Roblek V, Meško M, Krapež A. A complex view of industry 4.0. SAGE Open. 2016;**6**:1-11. DOI: 10.1177/2158244016653987

[18] Gorecky D, Schmitt M, Loskyll M. Human-machine-interaction in the Industry 4.0 Era. Proceedings–2014. In: 12th IEEE International Conference on Industrial Informatics, INDIN 2014; 2014. pp. 289-294

[19] Palma J, Bueno U, Storolli W, Schiavuzzo P, Cesar F, Makiya I. Os princípios da Indústria 4.0 e os impactos na sustentabilidade da cadeia de valor empresarial. In: 6th International Workshop–Advances in Cleaner Production. 24th to 26th May. São Paulo. Brazil; 2017. pp. 1-8

[20] Schlick J, Stephan P, Loskyll M, Lappe D. Industrie 4.0 in der praktischen Anwendung. In: Bauernhansl, Industrie 4.0 in Produktion, Automatisierung und Logistik: Anwendung, Technologien und Migration; 2014. pp. 57-84

Chapter 2

Industry 4.0: Current Status and Future Trends

Jesús Hamilton Ortiz, William Gutierrez Marroquin and Leonardo Zambrano Cifuentes

Abstract

This chapter focuses on the vision of one of the biggest changes that will mark our way to live: Industries 4.0. Our intention is to discuss its basics, the automation and improvements on the processes, the way to transform an small and medium enterprises (SMEs) into an Industry 4.0, some of the economic and educational aspects, return of the investment, etc. This topic is very extensive; nevertheless, our purpose is to give a current vision of the industry and also show the Industry 4.0's future trends. Furthermore, this chapter presents the process to migrate to Industry 4.0 until reaching to Society 5.0 and, in turn, includes a future version of Industry 5.0 that is expected to begin in 2020.

Keywords: Smart Factory, digital factory, pillars, economy, SME, migrate to Industry 4.0, Society 5.0

1. Introduction

The fourth industrial revolution happened as an evolution process in terms of technological advances, following the trend of the first industrial revolution; this first industrial revolution occurred in Britain, during the eighteenth century; it represents the social, economic, and technological changes that affected all Europe, and with the pass of time, were going to affect North America too [1].

The industrial revolution meant the end of the manual labors, as also finished with the use of animals for traction tasks; at the same time, new ways of work were developed, which reduced the working hours without affecting the amount of production. It is important to have in mind that this new instauration of order was not happening in a fast way and it required so many years to obtain a notorious process, which meant something more than a revolution. It was an evolutionary change that affected not only the industrial world but also the economics, population, the social structure, the culture, the relevance of some institutions, and, in a general way, the daily life. The industrial revolution transformed completely a nation full of farmers and craftsmen into the workforce nation, with some new revolutionary machines (steam machine and knitting machine, among other inventions). The first industrial revolution had a great impact in the textile industry [1]. This important change of order started around the 1750s and ended during the 1850s.

Some of the most important characteristics of the first industrial revolution were: the use of the motor force in the industry, the development of the manufacturing system, and the speed improvement in transport and communications, among others. The development of machines in each of the work fields in Europe during the middle ages meant a positive and an important improvement which helped to develop the first industrial revolution. One of the most important machines created during this time was the hydraulic mill, used to grind the wheat. This specific mill had the same system in the monasteries located along the majority of Europe lands, like Portugal, Scotland, Sweden, and the Czech Republic [2].

In England, at the end of the eleventh century, there was a huge hydraulic power that expanded almost 34 ropes and "there was 5624 mills and 3463 castles, which mean that more than the third part of all these castles had at least one mill, even two" [2, p. 184].

The middle age can be considered as one of the relevant moments in our history, where the fundamentals of the modern capitalist economy were settled.

Some of the biggest factors which boosted the economy and the first industrial revolution on the later days in Europe were the geographical discoveries in South America (during the colonization period), which provided new agricultural sources, not to mention the wealth increase; all of these background contributed to the capital accumulation [2]. Also, the monarchical state guaranteed the transport and was protecting all the goods. It was in charge to collect the taxes, and decided who could access to get money loans, which made easier to create great business. The renaissance and the religious reformation ended up promoting the search for the progress; in addition, the humanism which characterized these movements promoted the progress into the society. Finally, it is possible to say that the facts occurred during the middle age in England were the prelude for the industrial revolution [2].

The second industrial revolution occurred in the United States. The innovative technology was the electric energy. It happened during the 1850–1870s period, until the first world war in 1914 [3]. In this new industrial revolution, some of the most important and biggest changes in terms of technology and science appeared; also, the revolution expanded its developments throughout the world. The innovative technologies were based basically on the newest energy sources, such as fossil fuels, the use of new materials to improve machines (like iron), and new transport and communication systems (railroads and steamboats). All these developments generated changes related to work, education, scientific field, and mostly with the human consumption, which brought the mass production concept. So many countries joined into this process of industrialization, such as Germany, France, Japan, etc. Icons of this method such as Henry Ford, Orville y Wilbur Wright, and Nikola Tesla appeared in the book of our human history as great inventors who contributed to this second industrial revolution [3].

> *"The first industrial revolution increased the productivity for the Great Britain, while the second and the third industrial revolution created the global economy and the international division of work, which were responsible of the upper levels of life obtained around all Europe" [3, p. 56].*

The third industrial revolution focused on the development of digital technologies, a notorious improvement in the communication, the development of the Internet, the development of networks, and the new removable energies. These one appeared during the twentieth century, but it was not approved until June 2006, by the European Parliament. All computers, new equipment, communication systems, and, in a general idea, all the digital technologies are made up to improve and make themselves bigger and, even in some cases, have made substitutions for some jobs, relegating the human mind.

The third industrial revolution has done the transition between the analogical and the digital concept; it has invented the integrated circuits that have helped to reduce production costs. The use of the communication technologies and informatics has been indispensable, making possible the assisted use of the computer, the use of optical fiber, telecommunications, studies on the genetics field, and the development of lasers [4].

The fourth and last industrial revolution brought advice and forecast into the system. This new age of revolution is based on the creation of an integrative and collaborative environment, with the intention to add cyber physical systems and to consider the customer as part of the production process. On this fourth industrial revolution, the analysis and the test of the product are done before it goes to the market, not after, which bring up the opportunity to predict what is going to happen before it happens, improving the management of the investment that is made around the production. The main input that came up with the fourth industrial revolution was the setup of the Intel industries, where it is not just about data compilation, but keeping in mind the importance of the interaction between all these information, being allowed to connect it with different information sources and being able to create a controlled system with a high speed and flexible kind of production [4].

2. The Industry 4.0 pillars

2.1 Theoretical framework

The fourth industrial revolution is the last advancement of humanity; it has been promoted from technological advances that the population have in their hands nowadays. It is mainly based on the digitalization or automation of the factories through the use of the Internet of Things (IoT); also, it is paired up with cyber-physical systems. These systems (or pillars) will be the autonomous kind of-systems, with the capacity to make their own decisions, applying machine learning [5]; it will also collect data in real time, which will be analyzed and saved into the cloud. Industry 4.0 will allow to exploit pillars such as the Internet of Things (IoT), Big Data and data analytics, augmented reality (a virtual representation of the real world), cybersecurity, collaborative robots, additive manufacturing, cloud computing, artificial intelligence, and finally, 5G networks. In addition, Rojko [6] says, Industry 4.0 could be capable to decrease the production costs by 10–30%, logistic costs by 10–30%, and quality management cost by 10–20%." "The Intel industry products have implemented sensors (IoT) connected to the network, capturing and storing information constantly into the cloud (or a server), with the purpose of being processed later; this means the information is closely related to the pillars of Big Data and data analytics. Also, the IoT is related to the cybersecurity pillar because of the interaction between the devices connected to the Internet, communicating between it selves constantly, making them highly vulnerable to a computer attack.

1. Collaborative robots, which are going to work with humans in the industry, making a significant amount of processes in an efficient way, are more sophisticated than their predecessors; these robots will allow to obtain a considerable decrease of costs related to the building of fences or safety cells that, in

the previous days, kept the robots isolated from the humans. Rounded-look alike limb from the sensitive robots has been investigated too, with the intention of avoiding dangerous blows that could hurt the operators; these robots will use high-quality strength sensors to guarantee more safeties to work with them. This new system makes possible for the robot to have the ability to stand up with a determinate strength, allowing him to avoid obstacles [6, p. 82].

2. Augmented reality tries to incorporate, or to make a mix-up with virtual elements, using some of the digital content as the standard tool to our tangible reality, in real time. The main idea is to add some digital information to regular and common objects that are observed; it ends up being extremely useful to understand in a deeper way these objects. This process is achieved using a digital system related to a camera, a GPS system, a 3D scale, and an algorithm which is in charge of making the bond with reality. These tools have the capacity to increase the human performance, providing the required information to make a specific task. Also, the collected information by the algorithm and the camera has to be situated on the right place, keeping in mind the respective object that is being observed. Augmented reality can reduce lots of bugs in an immediately way, reducing time that is usually used to repair it [6].

3. As far as cost reduction is concerned, the simulation pillar is the most indicated and outstanding. Simulation modeling is the method that makes use of real models or imagined system models, which allows to make some experiments to validate designs, processes, or systems [5].

Simulation modeling allows to know about the complex systems through the development of complex and versatile products, which allows testing new concepts and systems, resource policies, and new operations before their actual implementation, allowing information and knowledge to be collected without interfering with the operating system [5, p. 9].

System modeling allows to decrease costs as well as reducing the time-lapse of developing and increasing the quality of the product. The imitation of the entire environment of a manufacturing factory allows to make an analysis to know on what point of manufacturing some problems could appear which could delay the production of the product [7].

An Intelligent Factory is not only about the digitalization of the processes but also related to the efficient use of energy, optimization of the processes, and the reduction of the impacts on the environment. At the moment, the best way to achieve this in an immediately way is using simulation. Simulation processes are carried out with advanced software tools, which have great economic impacts on manufacturing factories. The visualization of these processes favors the economy of the factories, since it allows to anticipate situations that could generate losses, to detect the weak points, and to improve phases of the production before putting them into operation [5].

4. Additive manufacturing is an enabling technology, capable to help with the new products, new business models, and supply chains. A set of technologies that allows "3D printing" of physical objects builds up the term additive manufacturing. Unique products can be manufactured without the conventional surpluses, making it a great advantage [5]. A better definition could be "additive manufacturing creates complex parts from the scratch, constantly adding one layer at a time, based on a 3D CAD model" [8].

With the 3D printing and the additive manufacturing, the operator could be able to have customized designs for clients with previously unimaginable figures (incredible geometric designs), compared with the ones that have been made up with its predecessor technology. 3D printing personalizes the design of the part that is needed for the final assembly.

The main idea for each customer is: they can make an order at their liking, and Industry 4.0 is going to take care of the whole relevant process to get the final and personalized result. Also, the additive manufacturing would reduce the consumption of resources and costs: "greater customization without the need of additional tools or manufacturing costs, maximizing the use of the material, fostering a 'zero waste' motto [9].

5. The pillar system for integration can be of two types: vertical and horizontal.

Horizontal integration is related to the inter-company integrality and is the basis of a close and high-level collaboration between several companies, using information systems to enrich the product life cycle by creating an interconnected ecosystem within the same value creation network [5]. Vertical integration is a network manufacturing system, is related to the intra-business integrality, and is the basis for the exchange of information and collaboration between different levels of the hierarchy of the company, such as business planning, production programing, or management [5].

6. Cybersecurity is responsible for providing protection to the stored information, either on a device, or on the network. The system is capable to protect the information from threats like computer strikes, or even from physical kind of threats. It is necessary to know that any system in the world is 100% secure and there always be a risk that must be assumed. In addition to local security (cybersecurity), it is important to talk about physical security (biometrics, safes, steel doors) which is an extremely important method that helps the user to preserve the information in a data center, just to make an example. Cybersecurity has its focus on three key points: data confidentiality, data integrity, and availability of data or CID [10].

Computer threats can be a virus, Trojan, and malware, among others, and they can proceed from an USB device and also from a strange mail with false advertising content. On the other hand, physical threats could be associated to the use of superpower force to break into the place where information is stored, or if the user's information to enter into the storage system is robbed. In the same way, there are encrypted, boot, filed, or the overwrite viruses, among others. The most important thing on Industry 4.0 is to provide good and safety practices to factories, which can be reviewed on the ISO 27001. The industries presented on this document have a vital importance because with the interconnection of devices due to the IoT incorporation, industrial plants become more vulnerable [10]. The increase in connected devices means more possibilities of cyberattacks [5]. "The IoT must be built on the basis of secure communications at each point of the manufacturing processes and it has to guarantee the interoperability of security between facilities as basic elements of the supply value chain" [5, p. 14].

The high connectivity required by Industry 4.0 has introduced more open and easily accessible systems to the world, which have increased the possibility of new cyberattacks. Nowadays, it is common to see in the industrial control system that connectivity is based on TCP/IP and Ethernet, or even in the use of standardized wireless systems. "For a variety of industrial attacks, software-defined networks

(SDN) and virtualization of network functions (NFV) can facilitate automatic response for incidents" [5, p. 14].

SDN and NFV make possible an automatic response to the incisions, for a faster detection of system failures, and, for a temporal time-lapse, replace these affected systems with virtual implementations. SDN and NFV are technologies to improve these following aspects: (1) network visibility, (2) network capacities (allows network traffic flows with better management), and (3) deployment and control of network functions using software, instead of specific hardware dle-boxes [5].

7. The IoT pillar is mainly about connecting "things" or objects to the network. The idea of bringing sensors and actuators to the industry would allow capturing data on real time related to the manufacturing process of a product, as well as the behavior of the industrial environment the will be analyzed later by the Big Data pillar. Since it is a large volume of information, it will be stored in the cloud with the help of the cloud computing system [6].

As individually distinguishable by the real world, "things" can be anything like an object or a person. Nowadays, IoT is widely used, for example, for transportation, healthcare, or public services. Thing-to-thing, thing-to-human, and human-to-human form a network within IoT, connected to the Internet [5].

This network allows the exchange of information between devices; it is an interaction which goes beyond physically, and it is focused on the virtual and in the digital concept of things. The object carries sensors so that the user can interact both physically and digitally (virtual), improving the user experience. "For different purposes, the digitized information can be used to adjust production patterns with the use of the physical world and using sensor data" [5].

8. The cloud computing pillar refers to the possibility of offering certain services over the Internet or the network, to a customer. "Assante and others characterized cloud computing for small and medium enterprises (SMEs) as a common fund of resources with rapid elasticity and measured service, self-service on demand and wide access to the network" [5, p. 3].

The implementation of the cloud computing pillar has advantages related to the reduction of direct and indirect costs, eliminating IT infrastructure in the organization and optimizing resources in a dynamic way, which users usually consume, or related to the portability of any type of devices, connected to the Internet which can be accessed from anywhere in the world [5].

Each service has its own characteristics that differentiate it between others: for software as a service (SaaS), the idea focuses on having a software in the cloud, with an option to offer its services from the same server to all the customers, being able to interact with any kind of software, program, or license created by the company, which from now on is going to be called the supplier. The great advantage of this system lies in the ease of managing, maintaining, and updating it. In the case of the infrastructure as a service (IaaS), what the provider offers is the infrastructure, its facilities, its servers, and its bandwidth, so customers can develop high-performance jobs, which need a lot of equipment hardware, high-speed Internet, or large data storage that a normal person would hardly have. The user pays for its use, avoiding the spending of administration and maintenance of servers or large data storage [11].

PaaS service "platform as a service" offers a combination of the previous two services, hardware tools and the basis of its software, to make a new application.

Customers do not need to install any software at home; the provider offers the basis of the software, adding the infrastructure to run it. With the PaaS, users have the possibility to create applications without worrying about building the infrastructure, or even the server's maintenance. In addition to this, it is important to keep in mind that users or customers can make their service package as scalable as they wish, either by adding more resources, decreasing resources, or migrating to the next level of service, while the latter refers to capacity and ease of being able to advance from one level of IaaS to the next as PaaS [11].

9. Big Data refers to a large amount of data structured as unstructured and semi-structured data that is difficult to manage, analyze, and process, given the complexity of the management of these kind of data, the speed with which these databases grow, and the multiple sources which are producing it [12]. The important thing is not the huge amount of data, but what organizations can do with these data. When these data are analyzed by a company, business, or an organization, it allows these companies to make big decisions to subserve its economy. All the data generated by the new technologies are used for its analysis, such as weblogs, radio frequency identification (RFID), sensors built into devices, machinery, vehicles, Internet searches, social networks such as Facebook, laptops, smartphones and other mobile phones, GPS devices and call center records. Finally, Big Data must be combined with a relational (structured) database, so the management could increase its effectivity [5].

What makes Big Data so important is the user is allowed to generate an opinion or a point of reference to information which the company did not know they could have, the possibility to obtain external information about the company that they did not have before, such as comments (or "likes") in social networks, results of marketing campaigns among others that can make us understand if the products or services are being received by the public. All these well-organized data has its importance to the company to find problems and solve them before they can occur or find trends that they can use to favor the economy of the company [12].

The quality of the data is the big problem facing Big Data, due to what is known as the Four dimensions (4Vs): volume, speed, variety, and Value; other experts mention more dimensions to take into account.

According to the Big Data definitions investigated, unlike traditional data processing, the first suggestion to characterize Big Data was related in terms of volume, variety, and speed, also called the Three Vs. In order to continuously process large amounts of heterogeneous unstructured data collected in formats, such as video, audio, and text, among others, in addition, other dimensions have been attempted for better characterization, such as: veracity, vision, volatility, verification, validation, variability, and value [5, p. 7].

10. AI (or Artificial Intelligence) is nowadays considered separately from collaborative robots. AI is starting to being taken into account as another important pillar for Industry 4.0. AI is the use of computer programs to perform complex tasks. These programs are installed on robots of any type (whether they are sensitive or collaborative) to perform tasks or jobs, giving the possibility to robots to adapt into situations in a faster an efficient way. AI helps robots learn autonomously, act logically, and communicate with each other. The use of collaborating robots, at the same time AI I'd being used, conforms a fundamental tool for the optimal and efficient operation of production processes in the Smart Industry or Smart Factory [13].

Simulation and AI methods provided an improved solution for the robot arm's movement path, which saves 5% of time, which means a significant improvement in productivity [13].

Thanks to the sensors installed in the production systems, AI allows to capture the energy consumption of the individual machines, analyze the maintenance cycles, and optimize them in the next step. It can also indicate when the operating data is defective. As the amount of data increases, the system optimizes its efficiency and allows more accurate predictions [13].

11. "The 5G network comprises a software-defined architecture, which allows dynamic programming to provide separate layers for different applications. This will allow new and diverse cases of business use" [14, p. 150]. It is important to talk about 5G technologies or networks since this will be mainly the IoT advance engine, which will allow more mobility and will support the incredible growth of the number of devices connected to the network that had already been mentioned before. This network is 10–100 times faster, and also has better capacity to support real-time applications than 4G LTE. With speeds of 10 Gb/s, latency of less than 1 mS, supporting several billion applications and billions of 5G machines, will provide an always active user experience [14].

Some of the various applications that 5G will allow will be: "high levels of network reliability: electrical networks, industrial control, traffic, cyber health and smart city management, high volume of information—remote video surveillance, etc. and battery-powered sensor networks low cost/low energy" [14, p. 150].

Nowadays, combinations between fixed and wireless networks are needed for large IoT projects, but with 5G, it will be different since it will have a reliability of 99.999%, latency of 1 mS, low energy consumption that will remedy the deficiencies of the technologies of existing communication and flexibility due to standardization to handle the large number of IoT devices. With an exponential growth of the following technologies that will drive Industry 4.0—additive manufacturing or 3D printing, sensor technologies, nanotechnology, artificial intelligence, robotics and drones—there is no doubt that 5G will be of great help but it will be important to add that new technologies will appear and Industry 4.0 needs a communication network technology that satisfies the industry over time, heterogeneity, security, and protection of industries [14].

5G will allow manufacturers to automate end-to-end operations and configure or eliminate virtually new product lines or entire factories. With billions of sensors, machine-controlled robots, and autonomous logistics, all of them capable of communicating and operating remotely in real time through 5G, manufacturers can achieve huge productivity gains. 5G will be the platform that will allow growth and transformation in many industries, contributing directly to social and economic development [14].

5G technologies can play a key role in the integration of new technologies; the aforementioned technologies and the pillars of Industry 4.0 can offer a platform to interconnect machines, robots, processes, self-guided vehicles, and goods, among others. 5G can be seen as an additional pillar that will be responsible for interconnecting and communicating other pillars quickly, flexibly, and safely, providing the support for the massive growth of IoT in an intelligent industry. In an upcoming future, 5G will be indispensable in any manufacturing company that intends to migrate to intelligent industry [14].

As for the pillars that will be used to design and create the architecture, it is planned to make the first visit to the manufacturing company and based on what can be analyzed of the company, the pillars will also be selected.

3. Transforming a small and medium enterprises (SMEs) into Industry 4.0

According to the European Commission, an SME is a company that employs less than 250 employees and has an annual balance sheet that does not exceed 43 million euros. SMEs, as a general idea, have deficiencies in many aspects (limited financial resources, lack of organizational culture, investment, and research and development in technology, among others). If they are compared to multinationals, in addition to most existing, I-4.0 maturity models are more focused on large companies than on SMEs.

The level of development in which SMEs are located (the majority is considered at a level 0) to address migration to I-4.0 is not taken into account. By conducting a survey of SMEs, it is possible to find out to what level of development the company is in and which should initially be improved to get into the transformation into Industry 4.0. The main aspect to keep in mind will be that a maturity model of Intelligent Industry/Industry 4.0 focused on SMEs is needed that takes into account the level of development 0, in which most of them are located; it is also convenient that the company has included a self-assessment tool to know the state in which it is and the progress toward Industry 4.0; it should be taken into account that the change from level 0 to 1 will require more time, effort, and resources of the entire company [8].

Small and medium enterprises from now on must take into account this new industrial revolution that brings with it a technological revolution to improve the competitiveness of the company, as is the case of SEI Laser that since the beginning of the 1990s has been evolving technologically and receives Industry 4.0 less disruptively. The data collection of the actors present in the production chain helped SEI Laser to improve labor productivity that increased up to 40% and improved the technical capacity of the laser process, among others. Industry 4.0 will help to find solutions to improve any of the three main elements of the business model such as capturing value, offering value, and creating value [15].

The migration of an SME company to Industry 4.0 depends a lot of states in which the company is located (its cultural organization, its information systems, organizational structure, and resources), which is possible to be known through a survey of the SMEs that helps to collect information to know the status of the company and finally focus on Industry 4.0 [16].

3.1 Automation and improving of process

Nowadays, the changing market requires production processes with flexibility, adaptability, and agility in real time for any companies. Thus, the manufacturing automation or Smart Factory is the response to this issue. Automation or digitalization brings benefits, e.g., with virtualization process, the physical world can be linked virtually (virtual and simulation models). Another benefit is real-time capability; with sensors (IoT), it is possible to analyze data continuously to react to any changes in the environment in real time. The Internet of Things (IoT) allows connection of machines, people, and systems in enterprise, improving production process [17].

When a company invests in "Industry 4.0," it will change technological aspects, product development processes, marketing, logistics, manufacturing, after-sales services, and security [18]. Higher quality, flexible production, advanced planning and controlling with relevant, real-time data, rapid reaction to changes in demand, stock level, errors by to mention other advantages and improvements in the enterprise [17].

3.2 How to migrate?

It is possible to say that migrating to Industry 4.0 is quite complex and it is not just about implementing new technologies and making large investments in equipment, machinery, software, hardware, robots, and education [15]. It is a vision and a disruptive change in the way companies have been working so far. There is already reports from experienced companies that are living in the process of transformation toward Industry 4.0 gradually, since it is not advisable to change the whole structure completely because it can generate major disorders in the production process. There are recommendations for migration and a transitional process between the current productive model and where the company wants to go, which is a digitalized and intelligent industry [16].

One of the models was proposed by Jhon Paul Kotter. This chapter is going to summarize in eight steps of the proposed methodology: create the sense of urgency to migrate; form a coalition; create a vision—visual communication; eliminate obstacles; integrate others by giving them faculties; generate short-term results; deepen changes; and create institutional changes. On the other hand, it is important to mention that there are a number of requirements related to tell if an specific industry is an intelligent one, digitalized or Industry 4.0 [19].

Among the requirements, it is possible to mention: an interconnected production model between the components of the production chain and each of its processes, the digitalization of documents improving efficiency and optimizing resources, an automated production model, digitized and integrated from the management to the production chain, return to school all the staff that work in the company to be a proactive member in each of the departments that make up the company, a new management model optimizing machine-machine relationships machine, man-man, a change in cultural aspects traditionally managed in self-management, management, business culture, an increase in production, personalization of the product or service thinking of the end customer [15, 16].

The productive model, which was thought related to the personalized need of the final customer, a complete transformation of the business model, both at the level of supply and demand, new service platforms which optimize the value proposition, optimize the management of data and information, adapt or create new standards in the different elements that are part of Industry 4.0, redefine the concepts of quality, ensure the prestige of the company, increase cyber security of data and information, modify or adapt the laws of different countries to the new production model, the adaptation of international relations at the commercial level, measure the environmental impact generated by the new business model, to include concepts such as happiness, joy, well-being in the people linked to the production process.

3.3 Investment return

Investment return is a key element in business world. For this reason, when migrating or digitizing a company, it must be done in a personalized way and according to the model of each business. During this process, technologic aspects should not be the only thing to focus between the transformations to a new production model. It is very important to deepen in the benefits and efforts involved in migrating to a digitalized industry. In order to do it, a key process is to meticulously calculate the investment return in each aspect that is migrated; it can be technological, training management, process improvement, automation, etc. [20].

These calculations must be performed in each area to know exactly the benefits that could be obtained when performing the migration, and with that information, the company executives will have better tools to decide on how to perform the migration to Industry 4.0 or how it should be adjusted to obtain a better investment return. It must be estimated over time, either in the short, medium, and long term, and it is recommended to perform it in different phases, in order to maintain a continuous feedback throughout process; for SMEs, it is approximately 12 years, while for large companies, it is approximately 7 years.

In the transformation process, there are some recommended steps: analysis of the company current state, well-defined objectives for transformation, a work plan to achieve the objectives, calculation of investment return in every phase and area where the transformation will occur, and a detailed report to the company executives.

3.4 Taxes

In general, finance, industry, and economy ministries are making great efforts for the digitalization of many companies of different industries. To track this process, they have different platforms that allow them to identify the advance of each company and the challenges they face when transforming to Industry 4.0. One of the relevant aspects in the new industrial revolution will be the resource management by the different ministries. At this moment, each country handles them with different platforms and tools, and the efficiency level is different for every case [21].

Perhaps the resources management problem, received by the state through taxpayers, depends a lot on cultural, social, and political characteristics and control systems. The finance ministries are not exempt to adapt all their instruments and working system for a fully digitalized resource and administration management. The objective is to improve efficiency and control of tax collection; more intelligent platforms and management process are needed that can share information with different industries in real time and at every step of the production; it allows to constantly coordinate the financial management between suppliers and clients, and the information provided can be evaluated in a faster and more efficient way [21].

For this, the tax agency, as well as the industries, will have to digitize in order to have a better management of public resources, minimizing the possibilities of criminal acts and evasion, resulting from bad economic management and bad real time information management, it will lead to an increase in public collection and allow to improve the citizens quality of life.

3.5 Standardization

The standardization in the productive chain and the business model requires interfaces that allow all levels of the productive sector to handle a common language, for each element to be fundamental for the proper management of the business model; and not to be independent entities performing their function individually under a common objective. The idea is that everything is integrated and to achieve it, a new productive culture is required.

On the other hand, the traditional standards known as the ISO and IEC will have to evolve toward new standards according to the digitized and intelligent industry. In other words, technological facilitators will be needed to make possible an Industry 4.0. These technological facilitators must be present in every existing pillars and other fundamental technological entities, such as 5G mobile telephony and new management software (CRM, ERP, PLM, etc.); the integration between management, logistic production, and others software will be fundamental for the transition to Industry 4.0 and the decision-making of the company's management.

3.6 Economy

At the beginning of the fourth industrial revolution, all economic, financial, social, cultural, and environmental models will receive a great impact; they must adapt to this new revolution. One of the biggest challenges is the negative impact in the environment of the traditional production model; it affects human life but there is not enough awareness to start facing it. Perhaps people think that it affected others and not themselves.

The large volumes of plastics produced go to the ocean, food produced that end up thrown into the trash, computers, telephones, cars in the stock of companies due to overproduction exceeding the demand, excessive electricity consumption, excessive use of gas vehicles, poor policies to mitigate and control environmental pollution, excessive CO_2 emissions including computer equipment and computer romos [22].

For all these reasons, Industry 4.0 is an opportunity to modify all these superficially considered aspects of the traditional economic model. There must be awareness that the natural resources are limited and it cannot be allowed to continue destroying them as is being done so far. The new economic model must consider that the natural resources must be protected and the environmental impact must be minimized. It should be tried to unify two concepts: create wealth and wellness but simultaneously cause minimal environmental impact.

Similarly, one of the big bets for Industry 4.0 is the productivity increase generating greater wealth, but a consequence of that wealth must be greater equity in the resources distribution as a result of good management of the fiscal and tax control by the ministry. A serious mistake would be to focus on the overproduction and natural resources exploitation, and neglect the impact it has on nature. Some economists call it a circular economy, biophysical economics, and ecological economics. This economy is based on three principles: preserve natural capital by controlling natural and renewable resources, optimize the resources used more times, promote a more efficient system that eliminates external and internal aspects in the productive chain and particularly that has minimum environmental impact, design no waste products, elastic and flexible diversity, renewable energy increase, consider everything as a system, re-evaluate the costs and prices management [23].

3.7 Legal challenges in Industry 4.0

It is a well-known fact for people working in science and engineering that this field has evolved faster than the legal environment or the countries' constitutions; it is important to stand out that most of the countries' constitutions are from past centuries, and what jurists do is adapt the technological and scientific advances to the current law. This has been done throughout the history and that is why it is noticeable that science and the legal framework in general are asynchronous and outdated, causing difficulties in implementation and scientific development in real time. In addition, non-tech experts take the final decision on scientific issues and technological advances. For this reason, adaptation of the legal framework is necessary and should be constantly evolving in parallel with the sciences and technological advances.

On the other hand, a digitalized environment brings new challenges to customers; whether they are natural persons or companies, the way of work and relationships will be re-evaluated; they must adapt to the new needs and also protection, security, etc. of the information are important. In the same way, it will be necessary to work on values and ethics since the most important element in the fourth industrial revolution will be the information management, the regulation to achieve it, and the adaptation of all the legal systems to a fully digitalized world.

3.8 Future trends: Industry 5.0 or Society 5.0

Industry 5.0 makes a great change of perspective; the core of Society 5.0 focuses on people as fundamental axis of the production sector. Both production and marketing fields agree that beyond the focus of Industry 5.0 is the Society 5.0. In Society 5.0, the products or services offered will be customized to the customer needs. The intention is to reach a fusion between technological development and human beings, with the main objective of people and machines complimenting their activities, and not people being replaced by machines. The use of cobots and robots is a fundamental change for collaboration of repetitive, danger, and unsafe tasks. Furthermore, the humans work will be intellectual production, which means it will be necessary to be qualified to be proactive in this society model [24].

This new approach of man-machine interaction is expected to increase production and offer greater satisfaction to both the worker and the final customer, with customized products. Again, is important to emphasize that Industry 5.0 goes beyond just a production process, and looks for a Society 5.0, thought and made for people and cobots integrated.

One of the great contradictions with previous generations is the model focused on people. The automation of processes, the introduction of cobots and the evolution of technology allow people to develop new skills in the production process.

In this disruptive approach, in order to achieve an intelligent society education need to change from its traditional form, new tools, software and hardware are needed and must be integrated with cobots, and high technological training of people for Society 5.0. A huge difference between Industry 4.0 and Society 5.0 is Industry 4.0 has robotics and other technological pillars as the center of the industrial revolution, while in Society 5.0 technology, complements or collaborates with human's work. It is important to stand out that technological advances achieved in Industry 4.0, increase the efficiency and productivity of Society 5.0 by adding the focus on the productive process to creativity and craftsmanship of humans. A different vision is expected, another relevant aspect in this new society is the environment as a priority and the circular economy [24].

Also, it is important to mention that the personalization of products and services is done according to the real market requirement, in order to avoid oversupply and unsalable inventories, which happens in Industry 4.0; the main idea is to produce what the population really requires, leading to a minimum over costs and waste due to overproduction. This also reduces the CO_2 emissions and the environmental impact.

The vision is a society where intelligent industries, autonomous cars, intelligent cities, etc. are integrated, besides the actions against climate change is: minimize all negative issues in business world and management systems. A more supportive society, with greater cooperation between people, communities and countries, greater trust among people, minimum corruption in public management, a more equitable resources distribution can be an ideal thought; regarding to markets, the basis of the production process would be greater trust and credibility between clients and companies [25].

The Society 5.0 is expected to be more inclusive and environmentally friendly; and also to have a better individuality management and a simultaneous integration with the environment and the society. The general expectation is a more sustainable world where environmental, social, and economic impact issues are related and integrated.

3.9 Circular economy

The circular economy has begun to be implemented in Industry 4.0 and will be the fundamental element for the development in Society 5.0. In this economy, wasting, destroying, or throwing away are re-evaluated concepts and are changed for reuse and recycle; nothing will be wasted and everything not used is considered as a resource for the productive process; so repairing, adapting, and updating will be part of the daily life, not only for a few but for all the people in Society 5.0. Innovating, recycling, and preserving will be daily concepts for the people participating in this society.

4. Conclusions

This document shows a vision of the present and future of Industry 4.0. This chapter includes an introduction to the history of industrial revolutions, the pillars of Industry 4.0 such as systematic review, 5G networks, and artificial intelligence which are incorporated as other pillars because some experts consider them important in the Smart Factory. An economic approach and a return on investment approach are examined to see the benefits provided by the Smart Factory. In addition, this document provides an essential tool in the process of migration to Industry 4.0, the survey. Finally, it covers what Society 5.0 is and a future vision of Industry 5.0 that is expected to begin in 2020.

Author details

Jesús Hamilton Ortiz[1,2], William Gutierrez Marroquin[3] and Leonardo Zambrano Cifuentes[4*]

1 CEO Closemobile R&D, Spain

2 Universidad Nacional de Educación a Distancia (UNAD), Colombia

3 Servicio de Enseñanza Nacional de Aprendizaje (SENA), Colombia

4 University ICESI, Colombia

*Address all correspondence to: rafael.zambrano@correo.icesi.edu.co

References

[1] Flashes Magazine. The first industrial revolution. In: *Revue économique*, vol. 19, no. 2. Vol. 32. Flashes Magazine; 2017. pp. 54-58

[2] IONESCU IG. The first industrial revolution and general features of the world economy between the 16Th century and 1780. SEA - Practical Application of Science. 2018;**VI** (17 (2)):183-186

[3] Flashes Magazine. The second industrial revolution. The Econometrics Journal. 2017;**41**(161):4

[4] Duarte AYS, Sanches RA, Dedini FG. Assessment and technological forecasting in the textile industry: From first industrial revolution to the industry 4.0. Strategic Design Research Journal. 2018;**11**(3):193-202

[5] Alcácer V, Cruz-Machado V. Scanning the industry 4.0: A literature review on technologies for manufacturing systems. International Journal of Engineering, Science and Technology. 2019;**Xxxx**

[6] Rojko A. Industry 4.0 concept: Background and overview. 2017;**11**(5):77-91

[7] Mourtzis D, Doukas M, Bernidaki D. Simulation in manufacturing: Review and challenges. Procedia CIRP. 2014;**25**(C):213-229

[8] Mittal S, Khan MA, Romero D, Wuest T. A critical review of smart manufacturing & industry 4.0 maturity models: Implications for small and medium-sized enterprises (SMEs). Journal of Manufacturing Systems. 2018;**49**(November):194-214

[9] Tofail SAM, Koumoulos EP, Bandyopadhyay A, Bose S, O'Donoghue L, Charitidis C. Additive manufacturing: Scientific and technological challenges, market uptake and opportunities. Materials Today. 2018;**21**(1):22-37

[10] Isaca. *Cybersecurity fundamentals-study guide*. In: 2nd Ed. 2017

[11] Muhammad Alyas Shahid, "Cloud computing security models, architectures, issues and challenges: A survey," December 31, pp. 602-611, 2015.

[12] Vaidya S, Ambad P, Bhosle S. Industry 4.0 – A glimpse. Procedia Manufacturing. 2018;**20**:233-238

[13] Benotsmane R, Kov G. "Economic, social impacts and operation of smart factories in Industry 4.0 focusing on simulation and artificial intelligence of collaborating robots". 2019

[14] Rao SK, Prasad R. Impact of 5G technologies on smart city. Wireless Personal Communications. 2018;**100**(1):161-176

[15] Bernstein J et al. The Impact of the 4th Industrial Revolution; 2017. pp. 1-138

[16] Li D, Fast-Berglund Å, Paulin D. Current and future industry 4.0 capabilities for information and knowledge sharing: Case of two Swedish SMEs. International Journal of Advanced Manufacturing Technology. 2019

[17] Mohamed M. Challenges and benefits of industry 4.0: An overview. International Journal of Supply and Operations Management. 2018;**5**(3):256-265

[18] Nunes ML, Pereira AC, Alves AC. Smart products development approaches for industry 4.0. Procedia Manufacturing. 2017;**13**:1215-1222

[19] Kotter JP. Leading change: Why transformation efforts fail. Harvard Business Review. 2007

[20] Ching KV, Lumpur K. A case study of return on investment for multi-sites test handler in the semiconductor industry through theory of industry 4.0 ROI relativity. 2019;7(3):23-40

[21] Hamid NA, Hamzah FHA, Noor RM, Azali NM. Determinants of reinvestment allowance (RA) tax incentive utilization in embracing industry 4.0. Polish Journal of Management Studies. 2018;**18**(2):94-104

[22] Luthra S, Mangla SK. Evaluating challenges to industry 4.0 initiatives for supply chain sustainability in emerging economies. Process Safety and Environment Protection. 2018;**117**:168-179

[23] Rajput S, Singh SP. Connecting circular economy and industry 4.0. International Journal of Information Management. 2019;**49**(March):98-113

[24] Fukuyama M. Society 5.0: Aiming for a new human-centered society. Japan SPOTLIGHT. 2018;**27**(Society 5.0): 47-50

[25] Savanevičienė A, Statnickė G, Vaitkevičius S. Individual innovativeness of different generations in the context of the forthcoming society 5.0 in Lithuania. The Engineering Economist. 2019;**30**(2):211-222

Chapter 3

Industry 4.0 Technologies: What Is Your Potential for Environmental Management?

Ana Claudia Pereira Carvalho, Ana Paula Pereira Carvalho and Núbia Gabriela Pereira Carvalho

Abstract

The Industry 4.0 is a new industrial model characterized by excellent productive, procedural, and environmental performance. This new industry is perceived as a truly sustainable manufacturing model. And one of the aspects that most contributes to this insight is the extensive set of largely innovative technologies in the industry 4.0. Largely because such technologies are not necessarily unheard of, what changes is the integration between them in the context of industry 4.0. From this explanatory context, the purpose of this chapter is to present the full potential of industry's 4.0 leading technologies for their highly effective Environmental Management. Thus, it is possible to understand how integrated technologies cooperate together for an environmentally sound and sustainable positioning of industry 4.0 in all its areas. The industry 4.0 has a more conscious use of natural resources, less waste, processes are leaner and the life cycles of its machines and equipment are longer. All these factors together with the technologies, contribute for the Environmental Management 4.0 to be disruptive in relation to the still industries 3.0. The chapter also presents the main challenges for the industry's 4.0 environmentally sound and sustainable performance through its technologies.

Keywords: Industry 4.0, sustainable manufacturing, advanced manufacturing, Environmental management 4.0, Technologies 4.0

1. Introduction

The Industry 4.0 or Fourth Industrial Revolution is a new and imminent industrial model characterized as advanced manufacturing. What most differentiates this new industry model from the three previous industrial revolutions is the integration of its components. These are represented by people and their human work, machines and equipment, and technologies. Industry 4.0's internal environment is more integrated, agile, accurate, synchronized, and encourages external stimuli to arrive and be received in the same way.

This new industrial model is also known for its high technological, virtual and digital positioning. Existing technologies in industry 4.0 are not necessarily new, what changes is the integration between them, allowing the industry to behave dynamically and quickly respond to internal and external needs. In the context of production, these technologies enable production processes to be more efficient, lean and faster.

In industry 4.0 as well as in today's industries, care for the environment, proper and conscious use of natural resources, optimization of physical resources used, waste generation and reuse, and the search for no waste must also be strictly observed. Industry 4.0 Environmental Management is as relevant an area as all of its innovative potential, as it must set guidelines, supervise actions and monitor processes so that they are consistent with an environmentally sound and truly sustainable industrial profile.

The aforementioned industry 4.0 technologies are strong supports for Environmental Management 4.0. The technological potential of the new industrial model is one of the key aspects that allow industry 4.0 to be known as a truly sustainable manufacturing model. Its technologies working in an integrated way, allows not only the environmental management process to be more solid, but the industrial activity itself in all its processes. These become more virtualized, lean, accurate and made on demand.

From this explanatory context, the purpose of this chapter is to present the potential of the industry's 4.0 leading technologies for their respective Environmental Management. For this, five next sections are distributed in such a way that the first one refers to the presentation of industry 4.0, the second one refers to the presentation of their respective technologies, while the third one refers to environmental management in industry 4.0, the fourth refers to the potential of technologies for environmental management 4.0, and lastly, the fifth refers to the main challenges for environmental and sustainable beneficiation through the use of technologies 4.0.

2. Industry 4.0

The term Industry 4.0 was first mentioned in the German language during the "Hannover Fair" event in 2011 in Germany, and came up as a proposal for the development of a new concept of German economic policy that is based on high technology strategies, which symbolize the beginning of the so-called Fourth Industrial Revolution [1, 2]. Thus, as of 2011, scientific publications on industry 4.0 began to emerge, increasing exponentially around 2013. And many scholars, researchers and business professionals have paid attention to the new industrial model and its potential technologies.

The industry 4.0 is an imminent advanced manufacturing model characterized by a comprehensive set of technologies that streamline industrial performance, making it more integrated, virtual, digital, with excellent response time to internal and external stimuli, and especially with exemplary environmental and sustainable behavior. The industry 4.0 itself includes a set of six design principles that intrinsically include technologies. Such principles are named Decentralization, Virtualization, Interoperability, Modularity, Real-Time Capability and Service Orientation.

One of the peculiarities of the industry 4.0 is its integration capability. The author [3] presents the fundamental and interdependent presence of two value chains, horizontal and vertical. **Figure 1** presents the first value chain, the horizontal, together with their respective elements.

The digitization process is on its way to both horizontal and vertical value chains. The horizontal value chain digitization works to integrate and optimize the flow of information and goods from the customer through their own company to the supplier, and so the information comes back again. This process involves the

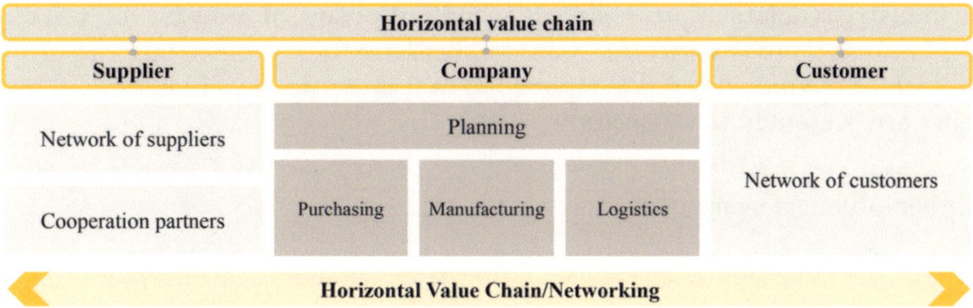

Figure 1.
Horizontal value chain in Industry 4.0. Source: adapted from [3].

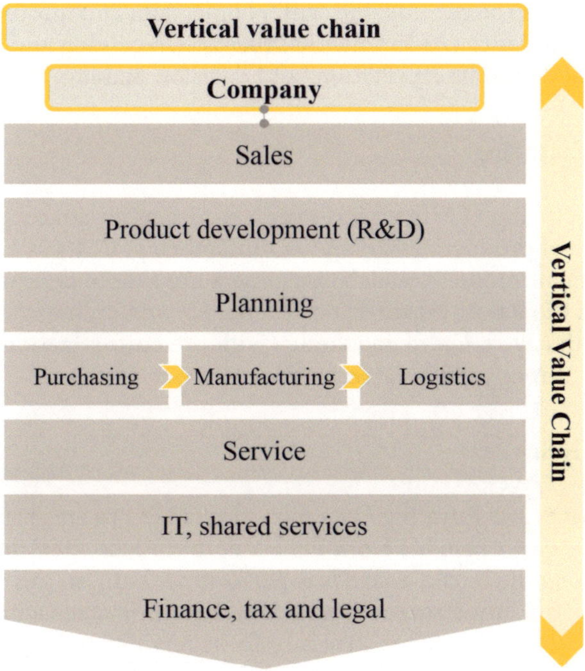

Figure 2.
Vertical value chain in Industry 4.0. Source: adapted from [3].

integration and proactive control of all internal areas of the company, such as planning, purchasing, production and logistics. The process also includes all external value chain partners who are key participants in meeting customer demand requirements and fulfilling the requested services [3]. **Figure 2** shows the vertical value chain with its respective members.

The digitizing the vertical value chain, on the other hand, is associated with ensuring a flow of sales data and information through product development for manufacturing and logistics. The quality and flexibility can be increased and costs reduced through the precise and optimal connection of manufacturing systems and their respective failure prevention, thus cooperating with better analytical skills [3].

3. Industry 4.0 technologies

The industry 4.0 itself encompasses an extensive set of technologies with broad innovative potential due to their integration. Such technologies are perceived by some industry scholars or professionals as tools or pillars, or industry 4.0 principles. This is because there is a need for more publications to be done in order to properly name each of the industry 4.0 components. In this section, key industry 4.0 technologies are presented in subsections.

3.1 Cyber-physical systems

One of the most significant advances in the development of information and communication technologies is represented by Cyber-Physical Systems (CPS). These are computational collaboration systems that are in strong connection with the surrounding physical world and its respective active processes while providing and utilizing data access services as well as data processing available on the Internet [4]. The CPS integrate the relationship with people, data and information, machines and equipment, in order to streamline the process of receiving and sharing data and information, analysis and interpretation, and decision making.

3.2 Cloud manufacturing

Cloud manufacturing (CMfg) is a research field that integrates cloud computing (CC) and traditional product design and manufacturing. CMfg is a service-oriented product development model in which consumers are able to design and make products through the use of information technology and online manufacturing resources [5]. In this way, CMfg cooperates with the customization of products made on demand directly deliberated by the customer.

3.3 Big Data analytics

Some scholars use the term Big Data analytics (BDA) to emphasize the process and tools used to extract insights from Big Data. In essence, BDA includes not only the organization on which data analysis is performed, but also the elements that make up the tools, the infrastructure and the means to visualize and present insights [6]. Thus BDA presents a stronger structure for receiving, archiving, analyzing and sharing data and information, enabling more concrete decision making based on real information and properly organized and treated.

3.4 Augmented reality

The augmented reality (AR) is defined as a computer graphics technique that transforms the real environment into a digital environment also using virtual objects in the real world [7]. The AR has several possibilities of use in different areas, either in the transmission of knowledge, in the performance of work activities, in the elaboration of products, in the deliberation of logistic routes, among many other applications.

3.5 Smart sensors

The Smart Sensors (SS) are one of the key elements of the future smart grid, as it enables remote monitoring at each specific point on a network to assess real-time system performance and to find likely errors [8, 9]. This technology enables the industry to better respond to internal and external needs for adaptation or change at a given point in the network.

3.6 Location detection

The location detection systems are designed to make it easier to identify the respective location of a user or a particular object in a physical space. However, the systems come in different varieties, some requiring users to have tags with identification that is viewed through fixed and calibrated fixed sensors. There are systems that provide object tracking anonymously, while some adopt fixed infrastructure (WiFi or Bluetooth). Location detection systems differ in aspects such as extent, area coverage, accuracy and cost [10]. While these systems differ at these points, regardless of the breadth of their potential, they enable the industry to have greater control over its resources and processes so that they are always optimized and directed to the right location.

3.7 Industrial Internet of Things

The industrial Internet of Things (IIoT) is a system that involves intelligent networking, cyber-physical systems, cloud computing platform that allows access, collection and evaluation of communications, and also enables the industry to exchange data, such as processes, products and services, in real-time, thus generating the optimization of production value [11]. Thus, IIoT is also one of the key technologies responsible for ensuring the integration of data and communications across the industry and beyond, in constant exchange of information with the stakeholders involved.

3.8 Additive manufacturing

The additive manufacturing (AM) is a technique that aims to bring together a wide variety of complex geometries and structures from three-dimensional (3D) model data. The process consists of printing successive layers of materials that are formed on top of each other. The AM involves various methods, materials and equipment, and has evolved over the decades and has the ability to transform manufacturing and logistics processes. The AM has been widely applied in a variety of industries including biomechanics, prototyping and construction [12]. This technology uses different materials for different purposes, for which aspects such as capacity, durability, consistency, limitations, and contact with other possible types of materials must be observed.

4. Environmental Management in Industry 4.0

The pace of technological progress opens space for reflection on its impacts not only on the economy, but also on society and the environment. Beginning with the Stockholm Conference in 1972, several meetings are held around the world to discuss environmental issues. Over the years there is a clash between the perspectives of industrial production, economic expansion and sustainable development. In addition to the use of water and energy, extraction of raw materials and exploitation of soil, there is a huge amount of waste generated [13].

The Environmental Management of traditional industries or industries 3.0 is very effective in its guidelines, supervisions and actions. However, traditional industry models do not allow complete and satisfactory environmentally sound and sustainable industry performance. These still have their processes that use many natural resources, which are sometimes over-harvested, the level of waste can still be considered high, the reuse of production resources does not occur effectively, as well as the reverse logistics process, among many other points that do not allow the industry to have a sustainable behavior.

The organizations seeking to perform well in sustainability actions must be ready to present the results of their practices to society and the market. To this end, the main mechanism used by corporations is the annual sustainability report. Through indicators, sustainability reports present the results of companies in the environmental, social and economic dimensions in a given year. This type of document contributes to broadening communication and relationships about these dimensions among key participants in the business environment in which the company is a part [14]. **Table 1** presents the aspects of environmental performance in the sustainability indicator proposed by [15].

In addition to the environmental performance aspects of the author's sustainability indicator [15], those proposed by [16] in **Table 2** are presented.

It is necessary to reflect on past actions and consequences and establish new relationships with the environment [10]. One of the goals set in 2015 by the United Nations (UN) in the 2030 Agenda for Sustainable Development is to promote modernization, including in industries with the aim of making them inclusive and sustainable with an increasingly efficient use of resources and implementation of technologies in industrial processes so that they become cleaner [17].

In this positive context, Environmental Management in industry 4.0 can count on its excellent ability to establish plans, deliberate on projects, grant or withdraw permits, verify processes and to monitor results, it can still use a wide range of integrated technologies that cooperate together for exemplary environmental and sustainable performance, which still has lower costs and expenses and increased industrial profitability.

Aspects of environmental performance	
Water utilization	Decrease in total water consumption
	Increase in percentage of recycled water utilization
Material utilization	Decrease of material intensity
	Decrease in raw material utilization percentage
	Increased use of recycled/remanufactured/reused material
	Decrease in percentage use of hazardous materials
Energy utilization	Decrease in total energy consumption
	Increase in percentage of renewable energy utilization
	Increased percentage of energy saved
Waste	Decrease in total waste generated
	Increased level of recyclable/remanufactured/reusable waste
	Decrease in percentage of waste brought to landfills
	Decrease in hazardous material waste percentage
	Decrease in water waste percentage
Emission	CO_2 emission reduction
	Decrease in greenhouse gas emissions
Source: adapted from [15].	

Table 1.
Aspects of environmental performance.

Aspects of the environmental dimension	
Natural resources	Water use (m^3/year)
	Recycled water rate (%)
	Land use (m^2)
Environmental legislation and compliance	Environmental accidents (No./year)
	Cost for environmental, health and safety compliance ($/year)
Energy	Energy utilization (kWh)
	Idle energy loss rate (%)
	Renewable energy fraction (%)
Material	Material used per production unit (kg/piece or m^3/piece)
	Scrap rate (%)
	Process scrap rate (%)
	Quantity of material reused or recycled (%)
	Use of packaging material (kg/year or m^3/year)
	Use of additive processes (kg/year or m^3/year)
	Quantity of recycled process additives (%)
Waste and emissions	Waste (kg)
	Hazardous waste (kg)
	Emission of ozone-depleting substances (kg)
	Emission of greenhouse gases (kg)
	Emission of other environmental gases (kg)

Source: adapted from [16].

Table 2.
Aspects of the environmental dimension (continuation).

Since industry 4.0 has a strengthened environmental and sustainable awareness, along with the use of its potential technologies, the acquisition of natural resources and all energy efficiency is done according to need and demand, without exaggeration. Thus, spending on the acquisition of productive inputs tends to decrease or, at least, to control, since the right and necessary volume of a certain resource to be used in production is being acquired. As well as the operating costs of generators, central computers, virtual and digital devices, machines, equipment, they are also reduced as they are activated at the right time and as there is a need to remain active. The minimal waste generated can also cease to be a cost to industry if it is reused as a new material to be integrated into other products, or even as a raw material for the construction of tools that assist in the production, or being the fuel for it, for some specific types of industry.

In addition to reducing costs and expenses, the environmental and sustainable behavior of industry 4.0 can increase its profitability. Industries in the same segmentation can create a network of partnerships in which the acquired resources that will no longer be used in production are passed on to other industries that need them, rather than rendering them useless or lost in industrial storage, leading to expenses and losses. Similarly, the use of machines and equipment can be shared across industries 4.0. This consists in sharing the operating capacity of industries and is a way of providing services to each other so that the supplying industry has the benefit of keeping its machinery running and generating volatile revenue when it is not operating in the industry from which it operates. At the same time, the industry using the machinery maintains its productive activity without interruption or loss of demand.

The interconnection of value generation networks in industry 4.0 provides new opportunities for the consolidation of closed-loop product lifecycles and industrial symbiosis. The network interconnection enables efficient coordination of water, energy, material and product flows throughout product life cycles, as well as between different factories. The closed-loop product lifecycles help keep products in life cycles with various phases of use, with remanufacturing or reuse between them. The industrial symbiosis describes the cooperation between different factories to gain a competitive advantage by trading and exchanging water, energy, materials and products [18] as well as smart data at the local level [19].

The approach to sustainable product design in industry 4.0 focuses on designing closed product lifecycles, enabling specific product reuse and remanufacturing or applying cradle-to-cradle principles. Different design approaches also focus on seeking consumer well-being. These concepts can be supported by applying identification systems, such as to recover cores for remanufacturing, or by applying new services that are additional to the product to achieve a higher level of customer well-being [20].

The transformations presented by industry 4.0 revolutionize the forms of commercialization and promote behavioral changes in the consumption and production spheres, and mainly, can cooperate for the structuring and formation of more sustainable business models [19]. Constantly these new technologies impact sectors such as health, energy and construction, and are configured in a junction of intelligent processes that originate products and services whose goal is to provide comfort and quality of life associated with the management of existing resources [21].

5. Technology potential for Environmental Management 4.0

The industry 4.0 just like every business organization, and every individual themselves, needs to have a strong environmental awareness. The difference with this new industrial model is that it can utilize an available set of technologies that enable it to have more environmentally consistent, clean and efficient industrial performance. **Figure 3** presents industry 4.0 technologies, demonstrated by using

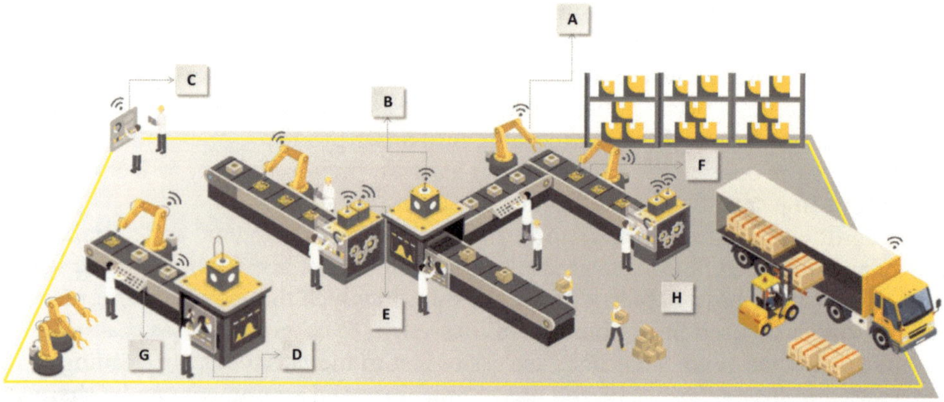

Figure 3.
Potential Technologies for Environmental Management in Industry 4.0. Source: Elaborated by the authors.

points, which have high potential favorable to Environmental Management 4.0. Each of the points is described subsequently.

5.1 Cyber-physical systems

The cyber-physical systems (CPS) are relevant agents for Environmental Management because they can receive production planning, stipulate production steps, provide data on resource utilization and production inputs, and manage them throughout the manufacturing process. The CPS can meet production schedules by receiving and providing real-time data and information, and can change any production step as needed.

Thus, the use of resources, mainly natural, is made more consciously and only in accordance with the need for integration in the production of the product. Similarly, less waste can be generated because the right materials are being used to the right extent and as needed. Because they are connected to the internet, and are the industry's leading interconnectors making it a network system, any changes that need to be made, the CPS can communicate it through notices to computer systems and mobile devices and perform autonomous (when programmed) or allowed (when coordinated by people), the necessary changes in the production process.

5.2 Cloud manufacturing

The cloud manufacturing (CMfg) is another potential agent for the consolidation of strong Environmental Management 4.0. This technology is responsible for receiving, storing, processing, presenting and sharing production data and information. Through this technology, not only the stimuli of the internal environment are treated, but also the external ones. Due to its service orientation feature, any minor change or need to adjust customer orders, notifications are sent and presented in systems to those responsible for a given production step.

The industry 4.0 has a high ability to offer customization to its customers' orders. By making it possible for customers to follow up the production steps of a product, they may or may not suggest modifications. This, in a traditional industry, could result in a vast amount of waste of resources, materials, energy efficiency, leading to waste and waste. However, with CMfg, it is possible to have the product planning to be executed, and thus acquire the necessary inputs. If any product change is requested by the customer, after systemic analysis, production can be changed without significant waste of resources, materials and efficiency, and other materials that can be used can be integrated into the production process, resulting in optimization of resources.

5.3 Big Data analytics

The Big Data analytics (BDA) stores and analyzes all industry data. Its potential for Environmental Management 4.0 is to group, analyze and provide relevant data and information on existing resources, resources used, machine functionality, energy efficiency, waste generation, waste utilization and pollution levels emitted. The BDA can provide historical information on a possible resource already used in the production of the same good, which may be reused. It can also notify, by systems, the lack of a certain resource for production, or even the breakdown of some machine, causing the production process to be interrupted.

With access to CMfg information, the BDA can suggest changes in the specificities of the production process. With these and other possibilities, the BDA serves as a warning of opportunities and correction needs, as well as its high capacity for storing data and information. Thus, any opportunity to improve production capacity by using fewer resources or taking advantage of fewer machines functionality, or by alerting to the misuse of a resource or material aimed at producing a good, or even the activity of a broken machine, can be notified by BDA, making it an environmentally active contributor.

5.4 Augmented reality

The augmented reality (AR) cooperates effectively with Environmental Management 4.0 because it can advise on planning before it comes to fruition, simulations of what is to be implemented, and the need for corrections even at a distance, or from different locations. It also contributes to the training and capacity building processes, imparting knowledge and teachings to employees about environmentally sound and sustainable actions through the visualization of real projections. It helps in the loading and unloading routes, so that truckers are better advised as to where they need to arrive, so that they are directed to the right location, without errors or deviations in the route and possible increase of pollutant emissions by their vehicles. The AR also cooperates with the monitoring of production operations, allowing to check failures, breaks, interruptions or errors in the production lines, among many other relevant functions.

With AR, plans can be prepared and viewed virtually before they are even directed, so any planning errors can be fixed before resources are acquired or triggered unnecessarily. The simulations that AR makes possible allow engineers, technicians, operators and designers to be able to check product-wide compliance and modification needs before they are produced. The AR still allows points that are distinct or distant from the industry site to be viewed virtually, so that the reality of a given point can be known even in real-time, thus streamlining the decision making process to perform actions or repairs.

5.5 Smart sensors

The smart sensors (SS) are a type of technology that informs systems for employee viewing of possible errors or non-conformities whether in materials, products, machines, steps or production lines, and can measure levels of waste or pollution generated.

The SS is a technology that can send these notifications to employees through systems on machines, central computers, or mobile devices and can take no further action, or can act, when so programmed, autonomously. The SS may receive an initial configuration of acceptable indices distributed at different scales, and when nonconformity that exceeds such ranges occurs, the SS may notify employees or take action that the production process is not erroneous at any stage, resulting in negative results for the environment.

5.6 Location detection

The location detection is a technology commonly used with the tag system. When you know the right location of machines and equipment, products in the production stage, finished products, a lot of time, accuracy, reduced waste generation and pollutants are gained throughout the industry.

The location detection helps resources that should be on the production line be easily located without wasting a lot of time and energy efficiency on running machines. Just as production-stage batches stored in temporary inventories do not need to be redone because they are not found, as well as completed batches need not be stopped in inventories generating industry costs and customer dissatisfaction, or even risk being confused with lots that have failed.

With location detection the industry becomes more agile and accurate, reducing waste generation due to local verification errors. Also logistics can prevent wrong deliveries and cargoes by mistake in defining the correct location to arrive.

5.7 Industrial Internet of Things

The industrial Internet of Things (IIoT) connects data and information from all areas of the industry and beyond with the stakeholders involved. The IIoT allows us to identify reversible or non-reversible machine and equipment failures, sensor and system problems, production disruption, changing customer demand requirements, stakeholder service opportunities, and many other possibilities.

Thus, the IIoT is a valuable technology for Environmental Management 4.0 because it can prevent the loss of a production machine or equipment because the hazardous situation has not been verified before, thus generating waste to the environment. Likewise, it can quickly verify that sensors and systems are not working properly and cannot perceive the improper use of a specific feature in a product, which could lead to loss of feature and product, leading to lost feature extraction and residue to the environment by the nonconforming product.

The IIoT also cooperates by informing customers of changing orders, so that the necessary resources and adjustments are handled carefully to avoid waste. Through IIoT, a network of industries can remain interconnected, so that with active communication, opportunities for offering and acquiring service claims can be observed and embraced, making it possible for an industry to need machinery that you do not own but can make available is to avoid buying machinery so that they will not be spare in this industry.

5.8 Additive manufacturing

The additive manufacturing (AM) is a technology that has a high degree of innovation in the use of resources in production. The AM's main contribution to Environmental Management 4.0 is that it can use other types of materials that are not necessarily taken from nature for the manufacture of various types of products.

With AM, some industry segments can switch production of their products using materials that do not come directly from nature, or that are biodegradable, or even edible, for some very specific types of industry. In this way the industry can achieve a more environmentally correct behavior regarding the acquisition of productive inputs.

In terms of waste generation, the AM can use materials that have been nonconforming or spare in the industry and make them viable to produce new products that integrate them rather than selecting materials to be extracted or prefabricated, eliminating irregular ones causing negative environmental results.

6. Challenges for environmental and sustainable beneficiation through technologies 4.0

The industry 4.0 technologies are very relevant to the environmental and sustainable commitment of industries. However, they do require a lot of observation and care in their correct implementation planning, operation, maintenance and discard. If all of these are not well analyzed prior to their realization, they can have an effect contrary to the environmentally correct and sustainable industrial positioning.

The Industry should be aware that technologies require continuous monitoring in order to ensure their full and effective use and no harm to the environment. **Table 3** presents the key challenges for the environmental and sustainable beneficiation of each of the industry's 4.0 technologies.

The industry 4.0 technologies are relevant contributors to environmental management, if properly and jointly used. One action of a given technology needs the cooperation of the other in order for efforts to complement and consolidate effectively.

Technologies 4.0	**Main challenges**
Cyber-physical systems	Cost considered still high for acquisition
	Installation in suitable places considering space and distance
	Need to connect to all machines throughout the production area
	Need to connect to all systems and sensors in the production area.
	Be connected in real-time to the whole productive area
	Pass by constant and periodic maintenance
Cloud manufacturing	Have ample data and information storage capacity
	Receive, archive and share data and information in real-time
	Be connected in real-time to industry computers, systems and machines
	Be connected in real-time with stakeholder demands
Big Data analytics	Have ample data and information storage capacity
	Receive, combine, analyze, archive and share data and information in real-time
	Be connected in real-time to industry computers, systems and machines
	Have historical information and records of changes made
	Be connected in real-time with the needs and changing demands of stakeholders
Augmented reality	Have interconnected auxiliary devices of adequate amplitude
	Have discipline in strictly necessary uses
	Be connected with systems and sensors for assisted production operations
	Have internet connectivity for use in different parts of the industry or in another location.
	Have internet connectivity for use in different parts of the industry or in another location that require direct interaction through communication
Smart sensors	Have space and distance required for correct performance
	Be connected to internet in real-time
	Be connected to systems in real-time
	Be connected to each other and machines in real-time
	Have the machines integrated performance capability
Location detection	Be properly and accurately installed in machinery and equipment
	Be connected to internet in real-time
	Be connected to systems in real-time
	Act together with smart sensors

Technologies 4.0	Main challenges
Industrial Internet of Things	Capacity installation on all machines, sensors and systems
	Be included on all central computers and mobile devices
	Receive, combine, share and notify industry internal and external stimuli
	Be included in partnering and service network systems from other industries
Additive manufacturing	Consider the best materials use option in the manufacture of a product
	Consider the need for resistance of the material to be used in the manufacture of the product
	Consider the use of waste generated as material for manufacturing
	Deliberate the ideal ways of disposing of the product when no longer used
	Control manufacturing according to need and demand, always considering the impacts on the environment

Source: Elaborated by the authors.

Table 3.
Main challenges for environmental and sustainable beneficiation through Technologies 4.0 (continuation).

7. Conclusion

In this chapter we have presented the industry 4.0 and its main technologies, so that its integrated structure and all the technological possibilities that the new industrial model includes are described. The context of Environmental Management in industry 4.0 was also presented, denoting the cooperation of the set of technologies 4.0 for environmentally sound and sustainable performance. Then, the potentials of each technology for Environmental Management 4.0 were presented, and finally, their respective challenges for environmental and sustainable beneficiation.

It can be noted that industry 4.0 is a model that uses the foundations of other industrial revolutions that occurred previously, but with greater innovation and integration. The Environmental Management 4.0 is disruptive because it can fully integrate the capabilities of the extensive set of technologies available to achieve systemic behavior that is more directed toward effective practices of consciously sourcing inputs, correct and accurate use as needed and demanded, less waste generation and reuse, quick perception of failures and corrections, real-time contact with stakeholders, among many other possibilities. However, as in traditional industries, excellent environmental and sustainable behavior requires awareness and collective action, and in the case of technologies 4.0, understanding and systemic actions.

Author details

Ana Claudia Pereira Carvalho[1*], Ana Paula Pereira Carvalho[1] and Núbia Gabriela Pereira Carvalho[2]

1 Federal University of São Carlos (UFSCar), São Carlos, Brazil

2 University of São Paulo (USP), São Carlos, Brazil

*Address all correspondence to: acpc@ufscar.br

References

[1] Lee J. Industry 4.0 in Big Data environment. German Harting Magazine. 2013;**26**:8-10

[2] Mosconi F, editor. The New European Industrial Policy: Global Competitiveness and the Manufacturing Renaissance. London, England: Routledge; 2015. 230p. DOI: 10.1430/84080:y:2016:i:2:p:367-369

[3] Koch V, Kuge S, Geissbauer R, Schrauf S. Industry 4.0—opportunities and challenges of the industrial internet. Strategy& - Formerly Booz & Company - PwC. 2014;**13**:1-52

[4] Monostori L, Kádár B, Bauernhansl T, Kondoh S, Kumara S, Reinhart G, et al. Cyber-physical systems in manufacturing. CIRP Annals - Manufacturing Technology. 2016;**65**: 621-641. DOI: 10.1016/j.cirp.2016.06.005

[5] Wu D, Rosen D, Schaefer D. Cloud-based design and manufacturing: Status and promise. In: Schaefer D, editor. Cloud-Based Design and Manufacturing (CBDM). Cham, Switzerland: Springer; 2014. pp. 1-24. DOI: 10.1007/978-3-319-07398-9_1

[6] Mikalef P, Pappas I, Krogstie J, Giannakos M. Big Data analytics capabilities: A systematic literature review and research agenda. Information Systems and e-Business Management. 2018;**16**:547-578. DOI: 10.1007/s10257-017-0362-y

[7] Ceruti A, Marzocca P, Liverani A, Bil C. Maintenance in aeronautics in an Industry 4.0 context: The role of augmented reality and additive manufacturing. Journal of Computational Design and Engineering;**6**:516-526. DOI: 10.1016/j.jcde.2019.02.001

[8] Morales-Velazquez L, Romero-Troncoso J, Herrera-Ruiz G, Morinigo-Sotelo D, Osornio-Rios R. Smart sensor network for power quality monitoring in electrical installations. Measurement: Journal of the International Measurement Confederation. 2017;**103**:133-142. DOI: 10.1016/j.measurement.2017.02.032

[9] Tang T, Ho A. A path-dependence perspective on the adoption of internet of things: Evidence from early adopters of smart and connected sensors in the United States. Government Information Quarterly. 2019;**36**:321-332. DOI: 10.1016/j.giq.2018.09.010

[10] Coyle L, Neely S, Nixon P, Quigley A. Sensor aggregation and integration in healthcare location based services. In: Pervasive Health Conference and Workshops; 29 November–1 December 2006. Innsbruck, Austria: IEEE; 2006. pp. 1-4

[11] Boyes H, Hallaq B, Cunningham J, Watson T. The industrial internet of things (IIoT): An analysis framework. Computers in Industry. 2018;**101**:1-12. DOI: 10.1016/j.compind.2018.04.015

[12] Ngo T, Kashani A, Imbalzano G, Nguyen K, Hui D. Additive manufacturing (3D printing): A review of materials, methods, applications and challenges. Composites Part B Engineering. 2018;**143**:172-196. DOI: 10.1016/j.compositesb.2018.02.012

[13] Fernandes I. A manufatura avançada sob a ótica da sustentabilidade: Estudo utilizando o enfoque meta-analítico. In: XXXVII Encontro Nacional de Engenharia de Produção; 10–13 October 2017. Joinville, Brasil: ABEPRO; 2017. pp. 1-22

[14] Nakagawa M. A sustentabilidade na estratégia de negócio das empresas brasileiras [thesis]. São Paulo: Pontifícia Universidade Católica de São Paulo; 2012

[15] Singh S, Olugu E, Musa S. Development of sustainable manufacturing performance evaluation expert system for small and medium enterprises. Procedia CIRP. 2016;**40**: 608-613. DOI: 10.1016/j.procir. 2016.01.142

[16] Winroth M, Almström P, Andersson C. Sustainable production indicators at factory level. Journal of Manufacturing Technology Management. 2016;**27**:842-873. DOI: 10.1108/JMTM-04-2016-0054

[17] Nações Unidas no Brasil (ONU BR). Transformando Nosso Mundo: A Agenda 2030 para o Desenvolvimento Sustentável [Internet]. 2015. Available from: https://nacoesunidas.org/pos2015/agenda2030/ [Accessed: 21 August 2019]

[18] Chertow M. Uncovering industrial symbiosis. Journal of Industrial Ecology. 2007;**11**:11-30. DOI: 10.1162/jiec. 2007.1110

[19] Stock T, Seliger G. Opportunities of sustainable manufacturing in Industry 4.0. Procedia CIRP. 2016;**40**: 536-541. DOI: 10.1016/j.procir. 2016.01.129

[20] Larsson T, Larsson A, Leifer L, Kobayashi H. Design for Wellbeing [Internet]. 2015. Available from: https://www.designforwellbeing.org [Accessed: 22 August 2019]

[21] Noronha M, Valente L, Rodrigues J, Bento L. Sustentabilidade 4.0. In: Congresso Internacional em Artes, Novas Tecnologias e Comunicação (CIANTEC 2018); 12–14 October 2018. São Paulo, Brasil: Pomello Digital; 2018. pp. 207-212

Chapter 4

A Taxonomy of Industry 4.0 and Related Technologies

Ercan Oztemel and Samet Gursev

Abstract

Industry 4.0 and related technologies will remain at the top agenda of manufacturing systems until respective digital transformation is completed. In order to increase the speed of the transformation and the respective performance, a taxonomy of industry 4.0 is proposed in this chapter. The taxonomy is defined through four aspects including strategic understanding, managerial practices, technological infrastructure and developments, as well as human intervention with respective skills and competencies. Each aspect of these is defined and further sub-categorized in order to reveal the real dynamics of industry 4.0 and respective implementations. Generating the taxonomy would also easy the categorization of the respective efforts and make the assessment processes to be carried out more effectively. It is also believed that the proposed taxonomy will be the source of generating a maturity model of industry 4.0. Note that the proposed taxonomy and respective components are defined by reviewing 620 papers, maturity models, and industry 4.0 projects.

Keywords: industry 4.0, taxonomy of industry 4.0, maturity models, industry 4.0 assessment

1. Introduction

Nowadays many technology providers are spending a great deal of effort for digital transformation through empowering the concept of industry 4.0. Especially, manufacturing companies around the world are now facing substantial challenges due to recent environmental, societal, economic, and technological developments. Aligning this rapidly changing world requires analysis in all areas, not just in manufacturing functions. It is therefore very important to understand scope as well as the structure and the philosophy of industry 4.0. Behind the need for a clear roadmap that will guide the transformation process, the enterprises should comprehend and prioritize basic components of industry 4.0 and identify the ways to cope with the respective challenges which in turn requires new capabilities for managing supplier value chain in an agile and responsive manner.

In order to allow close cooperation and swift adaption along the whole product/process lifecycles from innovation to production and distribution, the companies will need some competencies such as integrating virtual and physical structures (see, e.g., [1]). Manufacturing processes, on the other hand, will be utilizing very limited amount of human intervention. Self-decision-making will be a dominant feature of the implementation plans. Similar to these, several other issues need to be sorted out for having a beneficial transformation. In this respect, the taxonomy of the basic components of industry 4.0 would help designers to better facilitate the change process.

This chapter is devoted to provide such a taxonomy which would allow comprehensive analysis of industry 4.0 from different aspects such as strategic, managerial, technical, and human intervention point of views. As explained below, each view is defined and further sub-categorized to reveal real dynamics of industry 4.0 and respective implementations. Generating the taxonomy would also easy the categorization of the respective efforts and make the assessment processes to be carried out more effectively. This chapter, in this respect, is considered to be one of the main sources for generating the maturity models of the digital transformation.

Current performance in production technologies is mainly driven by the increasing efficiency and effectiveness of the respective manufacturing processes. There are various advancements at various levels such as:

- Organizational-economic level involving methodologies on lean management, process management and strategic management, change management, innovation management, knowledge management, etc.

- Manufacturing technology level including distributed manufacturing, virtual manufacturing, holonic manufacturing, agile manufacturing, additive manufacturing, robotics and laser technologies, etc.

- Material level covering semiconductors, nanomaterials, carbon fibers, thin films, biomaterials, etc.

- Information technology level, examples of which include RFID and embedded systems

All these achievements have led to major but isolated gains in process efficiency and product quality as digital integration of those creates new era of manufacturing. This also increases the possibility of generating intelligent systems enriched by smart sensors, the so-called cyber-physical systems. The integration takes place on both horizontal (across all participants in the entire value chain) and vertical (across all layers of automation) levels of a certain company as a whole. It is fostered by the following driving forces [2]:

- *Vision*: Industry 4.0 is a part of smart networked world, and the philosophy includes novel business, new social infrastructures, and real-time interactions. Transformation process would have to be triggered by a clear vision of the manufacturing. This vision is to be based on "product," "intelligence," "IT network," and "communication."

- *Strategic approach*: The new era of manufacturing is formed by two main strategies, mainly "leading supplier strategy" and "leading market strategy," aiming to be the pioneer in supplying and keeping a higher market position.

- *Requirements*: Companies should determine their needs by an in-depth analysis and see their strong and weak points to trigger the strategies accordingly. A comprehensive set of requirements will direct the implementation.

- *Priority areas*: A ranking should be made to strengthen the weak sides of the enterprise. Due to limited amount of resources, the company should generate a prioritized list of requirements and fulfill them accordingly.

Although there are some definition and foresights about the advancements toward future manufacturing systems or industry 4.0 operations, managing complex systems, delivering infrastructure for industry, safety and security factors, regularity framework, etc. are to be the main body of the possible roadmap for implementing practices outlined by industry 4.0. Keeping this in mind, a systematic definition of the future organizations helps both reorganize and advance the existing knowledge. This reorganization can be achieved through the classification of the present knowledge fields toward maturation. There are four main benefits expected as the following:

- Classification of a knowledge field provides a common terminology, which eases the knowledge sharing.

- Classification can provide a better understanding of the interrelationships between the knowledge artifacts in a certain field.

- Classification can help to identify knowledge gaps which can be subject to further improvements.

- Classification can support decision-making processes and facilitate generation of implementation plans.

Additionally, the classification of knowledge can support researchers and practitioners in generalizing, communicating, and applying the findings or advancements to the related fields.

The taxonomy on industry 4.0 presented here in this chapter is generated by reviewing 620 papers as outlined in **Table 1**. The authors benefited mainly from eight publication databases (CiteSeerX, ACM, AISeL, EBSCOhost, Emerald Insight, Taylor Francis, Science Direct, Google Academic) for both academic and business areas. Some other literature found on the Internet is also reviewed in accordance with their relevance. Note that the literature review carried out in this study is published in [3]. The methodology for generating the taxonomy is presented below. Following this, the current state of the art on taxonomy studies is given. The proposed industry 4.0 taxonomy is then explained in detail then. Note that the definition of industry 4.0 and respective components are intentionally not covered as the respective information is provided in various chapters of this book.

	1987-1990	1996-1999	1999-2002	2005-2008	2008-2011	2011-2014	2014-2018	Total
Cloud						27	25	52
CPS						51	43	94
Data Mining					5	5	34	44
ERP			2	10	5	10	13	40
Industry 4.0	1	1		13	30	31	37	113
IoT			1	9	14	1	23	48
M2M			10	25	2	10	17	64
Smart Factory			10	14	1	23	18	66
Virtual Eng.	1		1	3	5	4	14	28
Augmented Reality					1	10	10	21
Intelligent Robotics					10	10	10	30
Projects						10	10	20
Total	2	1	24	74	73	192	254	620

Table 1.
Number of literature reviewed for developing the proposed taxonomy.

2. Methodology used for generating the taxonomy

As stated above, the taxonomy is mainly a classification methodology. Enumerative and faceted approaches are used for classification. In enumerative approach, all classes are constant, making a classification plan intuitive and easy to apply. However, it is hard to enumerate all classes in immature or evolving domains. In faceted classification, views of classes that can be combined and extended are described. A true hierarchy ensures the mutual exclusivity, i.e., a presence can only belong to one class. In hierarchical classification, it is compulsory to have good knowledge on the subject or phenomena to be classified. The classes and differentiating criteria between classes must be well defined.

To generate a comprehensive taxonomy for industry 4.0, academic literature, popular media, websites, and Wikipedia as stated above in **Table 1** are first investigated. Attempts to search for respective work indicated that no taxonomy study similar to the one proposed here is available. The literature collected is carefully refined, and those contributing to the taxonomy is selected and put into a database for their contribution.

Similar to the literature review, some of the maturity models such as the one described by [4] and running projects such as MetamoFAB [5] are also reviewed. Together with the literature review, this analysis also showed that the structure of the industry 4.0 is shaped from different aspects, examples of those given in [6–10]. Those are categorized under the so-called common views as shown in **Figure 1**.

After the literature survey and analysis in this manner, attributes and components of industry 4.0 were identified and categorized. Following this, a taxonomy table was created with a wider focus that includes not only the components of industry 4.0 but also all related business functions. The proposed taxonomy is approved by a scientific committee where academics and industrial practitioners took part in this assessment.

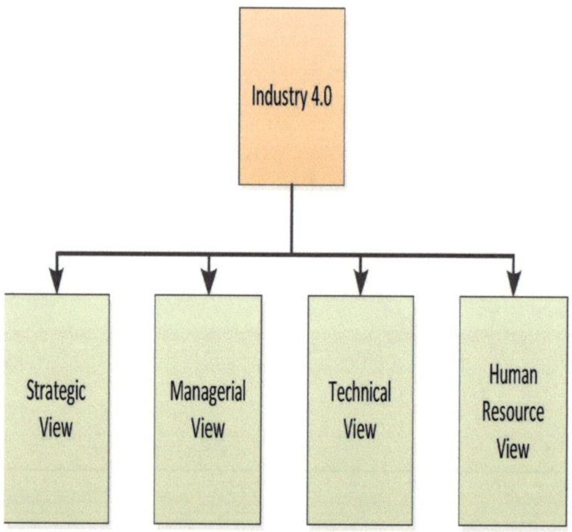

Figure 1.
Main views of industry 4.0, taken as baseline for the taxonomy.

3. Taxonomy of industry 4.0

According to Oxford Dictionary, taxonomy is "a scheme of classification." It allows for the description of terms and their relationships in the context of any knowledge area. The respective data within the domain is classified into a hierarchical structure based on their relatedness. As indicated in **Figure 1**, the taxonomy proposed in this chapter is based on four different views which could be considered as four pillars as explained below.

3.1 Strategic view

This indicates strategic aspects of industry 4.0 implementations, believing that the technology by itself is not enough to define the roadmap directing the transformation process as effectively. Some other factors such as strategic approaches should also be taken into account. Accenture investigated the main characteristics of industry 4.0 [11]. Klaus pointed out that new technologies like 3D printing, cloud, or CPS are important for industry 4.0 main goals [6]. But industry 4.0 also grows new strategies and new business models. This would enable to better foresee the future and generate respective assessment models. As shown in **Figure 2**, several sub-categories are defined by focusing on the strategic aspects of industry 4.0.

Corporate culture is an unavoidable part of any transformation in industry. It is elaborated in terms of knowledge sharing, open innovation, mission and vision, innovator and open innovation, etc. Total quality control is found to be important in order to utilize systems developed in an integrated manner. It is believed that these should be taken as the main drivers to restructure the organization. Especially mission and vision of the company should be re-addressed, and innovation culture should be given the prime importance.

Similarly, Kagermann et al. and all other studies explain the importance of ***roadmap for strategic evolvement*** [12]. Based on the literature analysis, the headings to be followed for the strategic roadmap of industry 4.0 were identified as the following:

- *Investments:* Industry 4.0 includes many new technologies which can be effectively utilized with the right investments empowering the competitive advantage.

- *Strategy:* Enterprise should be determining a correct strategy including industry 4.0 values as described by [13]. It is important to support especially innovation and be open to valuating new ideas.

- *Corporate approach*: Enterprises should develop implementation plan and a corporate approach for the activities to be carried out within and outside the organization. Pre-planned, specific rules, principles, or corporate culture should be established.

- *Resources for realization*: The diminishing amount of natural resources can be a problem especially for manufacturing companies with global supply chains. Industry 4.0 should try to enable the realization of the productivity growth that people need to satisfy the ever-growing consumption behavior, while not actually increasing the demand for resources significantly.

- *Implementation roadmap*: All departments and processes such as procurement, production, sales, marketing, and operations should be in line with the industry 4.0 standards, principles, and specifications which should be implemented in a roadmap manner.

- *Adaption of business models*: The business models in developed markets to generate competitive advantage are often challenged by the environment

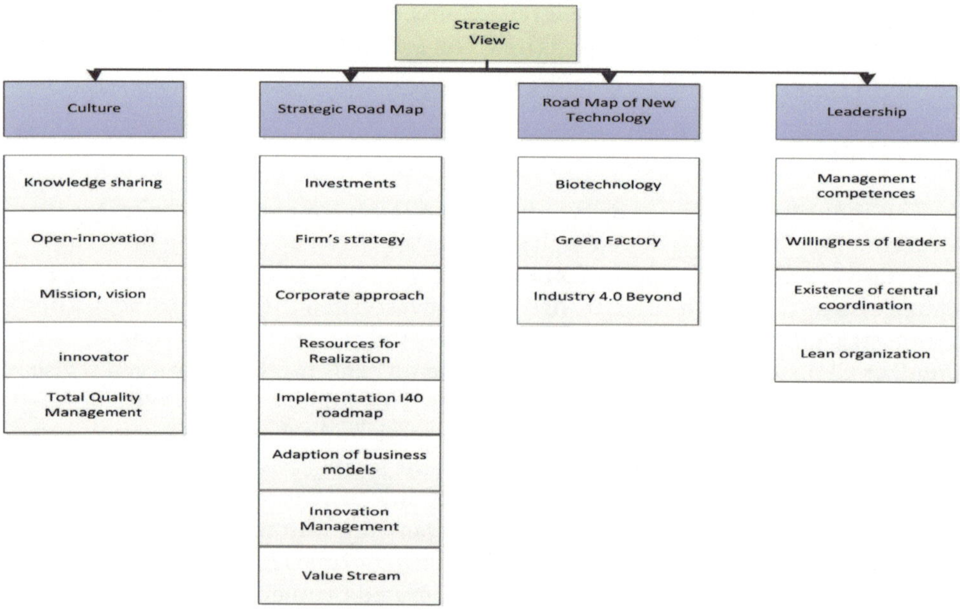

Figure 2.
Strategic View of Industry 4.0.

present in emerging markets. Achieving this change is important for digitizing the processes.

- *Innovation management*: Digital transformation is all about the change, research, and development operations with a systematic implementation process. Innovation management, idea development, conceptualization, feasibility, products, and service development, testing, launching, and marketing are the basic pillars of this process as they can provide their customers with a different perspectives and service understandings.

- *Value stream*: The set of all steps from the start of value creation until the delivery of the end result to customer should be the main focus of industry 4.0.

In the literature, it is hard to see all of these to be covered by any industry 4.0 implementation. Among those, investments, corporate strategy, corporate approach, and resource for realization are challenged in industry 4.0 projects. The rest of those are basically seen in reviews and maturity models (see, e.g., [14–17]).

A road map of new technologies is especially important for technology developing companies. Biotechnology and Green Factories are among those. A basic requirement for a "green factory" is energy-efficient and sustainable production that balances environmental and business interests. Biotechnology is also the technology that utilizes biological systems, living organisms, or part of this in order to develop or create different products. After industry 4.0, it is foreseen that the companies performing Research and Development (R&D) as well as making investments in this field will be successful.

Similarly, the ***leadership*** is considered to be an important criterion for strategy formulation within the taxonomy. It may include management competencies, willingness of leaders, and existence of central coordination as well as lean organizations. "Competencies" of managers surely can improve management skills and influence the behavior of others—which can have a positive impact on the bottom line.

The competency in this case is defined as the quality of being competent, adequate, and equipped with required skill, knowledge, qualification, or capacity. Organizations should define the competency requirements for their leadership team. This enables the organization to be managed with a consistent competency model. It is important for the leader to believe in the process and to be "willing" to digitize the organization. A leader must have three fundamental characteristics such as knowledge, courage, and compassion with a will to work. Note that in industry 4.0, central coordination and management are important requirements. Besides the leadership, there is also a need to control the whole enterprise and support the leadership with a good business model, especially where the entire factory is remotely controlled and managed.

A ***lean organization*** understands its customer value and focuses its key processes to continuously increase that. A structure other than lean organization cannot be accepted in competitive and innovative companies. A company with components of industry 4.0 is expected to be free of waste and inefficiency.

3.2 Managerial view

Managerial view is analyzed in terms of three main aspects as the following:

- Products
- Operations
- Business

The products to be produced can be in a variety of forms from ballpoint pens up to robots. However, the product should have some characteristics such as being intelligent enough to help its users to get the maximum benefit of it. Autonomy, self-behavior, being able to communicate with surroundings, generating useful knowledge for the users, etc. are some of the characteristics that have to be considered while transforming the enterprise toward the future. Similarly, industry 4.0 emerged to offer new approaches for dealing with complexity and improving productivity. By deploying the right combination of technologies, manufacturers can boost speed, efficiency, and coordination and even facilitate self-managing factory operations. Several research points out those as the basic benefits to gain in business (see, e.g., [17–19]). Keeping these three aspects into account, the components of managerial view is depicted in **Figure 3**.

As shown in **Figure 3**, the managerial view is categorized by managing the products, operations, and the business as a whole. From the ***product*** point of view, it is important to generate a *product integration scheme* which would mean something like a form of advertising as placing the specific product in a scene (see, e.g. [20, 21]). Note that, easy and fast integration of products into the market will be one of the main requirements.

While trying to digitize the processes, the way to *digitize the products* (being smarter and having information sharing capability) should also be considered. It should be noted that *big data* and *data analytics* are the main components of industry 4.0. This involves complex applications such as predictive models, statistical algorithms, and what-if analysis powered by high-performance analytic systems.

Similarly, *individualization* or *customization* is another unavoidable progress along digital transformation powered by industry 4.0. This could be reflected by the products as well as the processes. Having the products individualized or customized means generating completely customer-designed specifications which will be of great priority in the market for customer attraction.

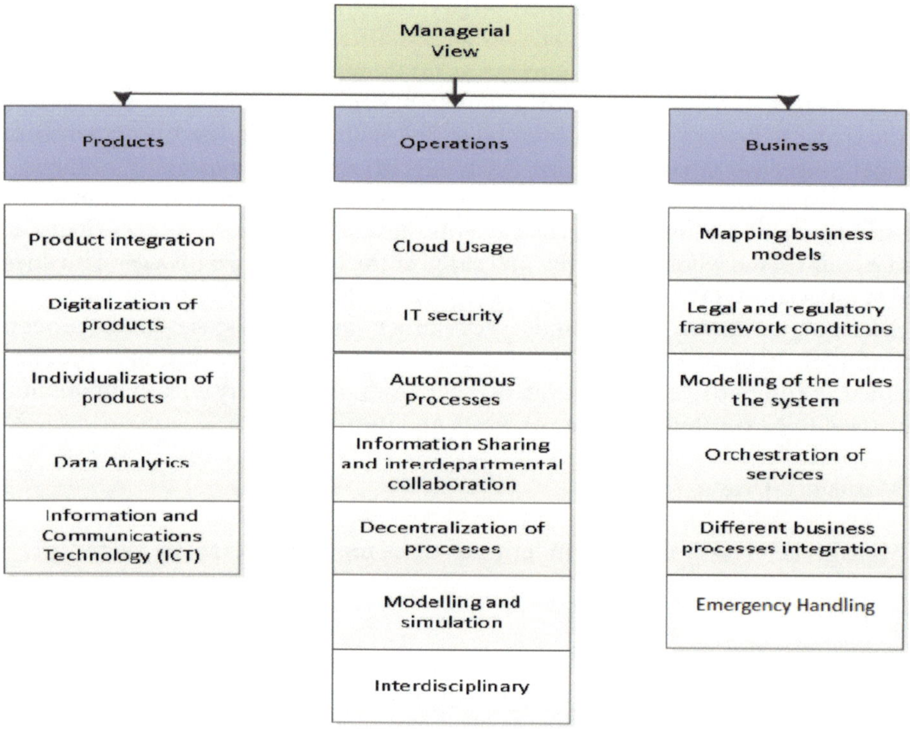

Figure 3.
Managerial View of Industry 4.0.

It is now obvious that industry 4.0 implies that information and communication technology will change human life. This can be considered as leverage for economic, societal, and interpersonal transactions and interactions. Modern information and communication technologies have created a global world, the so-called global village, in which not only the people can communicate with others across the world as if they were living next door but also machines, robots, equipment, and tools can be facilitated along this line. From the ***operations*** point of view, operations management of future manufacturing systems must have the following characteristics:

- Selecting an appropriate resource is mandatory for operation management. Using efficient process will eliminate the waste of efforts and redundancy. Today, *cloud computing* operations have become very popular and widely employed by many of the organizations just because it allows to perform all business operations over the Internet without spending too much time and money. This technology presents remarkable changes and opportunities especially in computing and communication.

- *Information security* is basically the practice of preventing unauthorized access, use, disclosure, disruption, modification, inspection, recording, or destruction of information. The field of information security has grown and evolved significantly in recent years. It offers many areas for specialization, including secure networks and allied infrastructure, and reliable applications and databases, etc. As foreseen by industry 4.0, future manufacturing systems should pay attention to the progress along this line.

- *Autonomous processes* is another expected behavior of manufacturing systems. This capability triggers the flexibility and adaptability ensuring subsystems and modules with a certain degree of decentralized autonomy and control. In order to manage dynamics, flexible, reconfigurable, and autonomous assembly systems were mainly realized in the last two decades and seem to be required more and more every day. Manufacturing systems taking industry 4.0 as baseline should heavily be engaged with these capabilities.

- Problems with *information sharing* and interdepartmental breaks while practicing industry 4.0 could be the main source of failure as Bello et al. clearly stated [22]. This implies a comprehensive setup of secure communication system. Since machines will be speaking to other machines and perform unmanned decisions, sharing information even knowledge between man to machine and machine to machine, as well as respective software, seem to be inevitable for successful transformation.

- *Decentralization* is the process by which the activities of an organization, particularly those regarding planning and decision-making, are distributed or delegated over an information network. No central system plays the role of control. The overall management of the operational processes is ensured by timely information exchange and requires quality.

- *Modeling and simulation* technology is one of the most popular components of industry 4.0 and revolutionized in the operational processes through augmented reality. As well known, modeling is used for multi-criteria analysis before the new product/process/operations are launched, and simulation is an important tool for the improvement of operation processes.

- Industry 4.0 requires *interdisciplinary operations* which facilitates value engineering, customer-oriented approaches, and customer loyalty. Enterprises aiming to experiment digital transformation should spend a great deal of effort on implementing these.

Similar to products and operations, industry 4.0 also reshapes the ***business***. The business model to be implemented should describe how the company creates, delivers, and captures the expected value. Business models within the scope of industry 4.0 include a large number of subcomponents as described by [23–25]. The proposed taxonomy includes the following elements:

- *Business model should be mapped* to the manufacturing operations. This mapping involves generating systematic solutions to target market/customer, channels of distribution, revenue streams, effectiveness of operations, intellectual property rights, customer relationship management, etc.

- The legal basis or *regulations* for which the companies are to obey is an important criterion for the system to align business requirements to regulations. That indirectly implies that the business model employed should be digitalized in accordance with the law.

- *Remodeling business rules* are necessary for deciding in which sequence the operations are to be executed. The traditional approach used to implement a business process in today's information systems will no longer be able to cover the actual needs of the dynamically changing business requirements. New form of rules needs to be developed especially for ensuring machine-software-human information exchange.

- *Orchestration* of services is another basic requirement of industry 4.0. This necessitates loosely coupling interoperable services of small units of software designed that perform discrete tasks. That is to say that the application or service orchestration is the process of integrating two or more applications and/or services together in order to automate a process or synchronize data in real time. Often, point-to-point integration is used. There has to be an integration plan to sustain required level of orchestration.

- Above the application integration, *Business Process Integration* (BPI) is also essential to enable automation of business processes, integration of systems and services, and the secure sharing of data across numerous applications. Overcoming integration challenges makes it easy for the organizations to connect systems internally and externally. Orchestration, process integration, and information sharing capabilities should be aligned to sustain the success of decentralization process.

Business continuity and emergency planning is necessary to assure the performance of industry 4.0 implementations. Since not much human intervention is necessary, it is required to get instant information from the business processes and control them especially under emergencies.

3.3 Technical view

The most detailed and most relevant characteristics of industry 4.0 are the technical aspects and technological approaches. The main reason for this is that the components are structures made up of new technologies. In order to identify related technical issues, the titles, keywords, and findings obtained in all projects, reports, and academic publications have been analyzed. After this extensive analysis, technical view of the proposed taxonomy is defined as shown in **Figure 4**.

The intelligent or **smart factory** is the most popular component of industry 4.0 (see [26–28]) with respective subcomponents.

- *M2M technology* is the main technical requirements for generating intelligent machines fostering smart factories. This technology allows interactions among machines, software, and human operators. Machine-to-machine communication is inevitable. Sensor or cyber-physical systems are used to enable this.

- *Optimization*, also known as mathematical programming, is the collection of mathematical principles and methods used for solving quantitative problems in many disciplines. Optimization is key component allowing information and material flow within the job floor and supporting the product development. The required level of flexibility is a must in order to be able to reconfigure the manufacturing suits for respective purpose.

- Industry 4.0 is all about running the *dynamic organizations* which should engage manufacturing resources, communities, and spaces for its employees and learners to thrive. Dynamic organization is considered as the primary key for the digitalization.

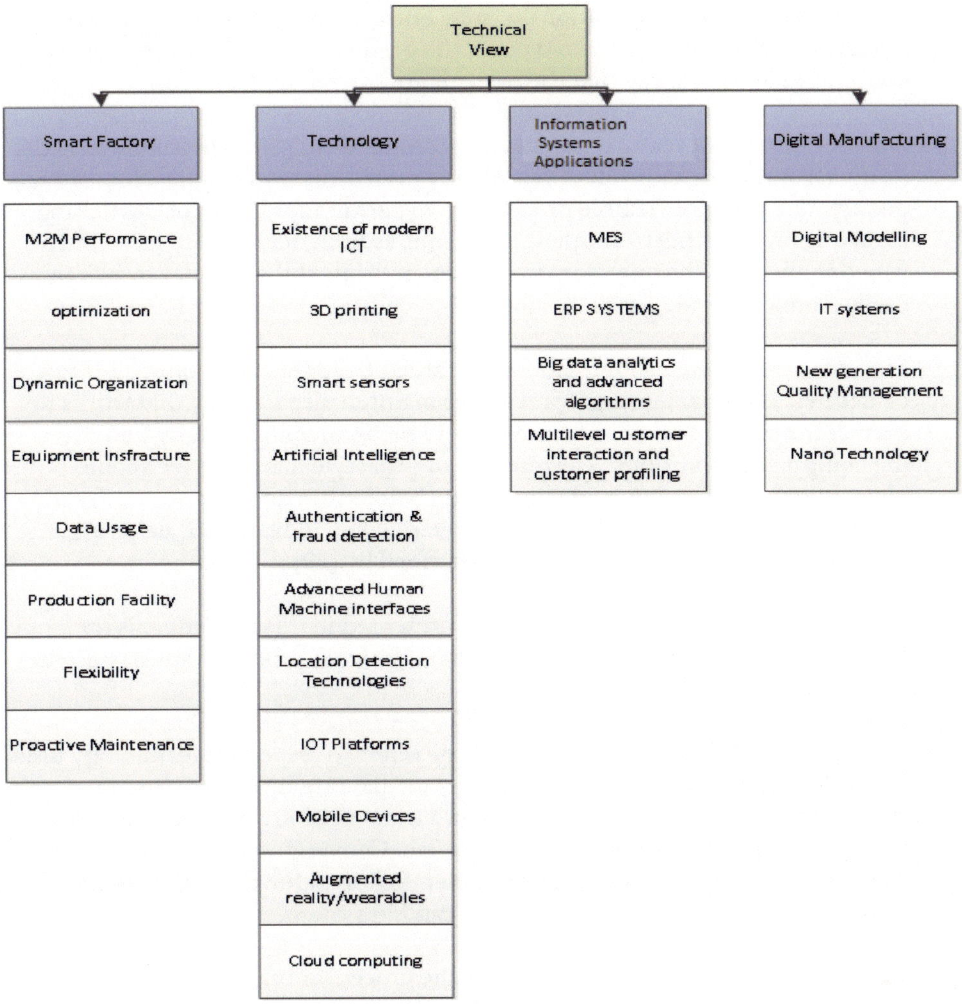

Figure 4.
Technical View of Industry 4.0.

- *Equipment infrastructure* is also important for ensuring business continuity and performance increase under dynamicity. Keeping interfaces set properly enable integrated functions to run without any shortcomings. Some of the applications such as equipment maintenance, optimization, and flexibility provide the link between the intelligent factory and industry 4.0. Machine maintenance is important as not much human involvement will be possible, and the machines are going to take control of themselves for continuously working.

- Similarly, the system would not be successful without *effective use of data*. Note that in industry 4.0, an extensive set of data is required to make the right decisions, understand and manage the processes, and perform improvement. Data is also necessary to define the required autonomy as it is an essential part of the process to provide sufficient capability for the enterprise to manage production facility for criteria of smart factory.

- *Production facilities* such as handling, separating, treating, and processing have to be fully under control of the system by itself. Respective equipment and facilities need to be in operation on time. Since assuring effective and efficient operations are a must for industry 4.0, the production facilities should be configured in such a way that the basic requirements are fulfilled.

- Another important feature to consider within industry 4.0 environment is the *flexibility*, meaning to produce reasonably priced customized products of high quality. The need for flexible processes is to permit rapid low-cost switching from one product line to another. This requires both machines to have multi-functionality and the operators to develop multiple skills in order to switch easily from one kind of task to another.

- *Proactive maintenance* should not be discarded from this taxonomy. This can be defined as implementing a preventive maintenance strategy that works to correct the root causes of failure and avoid breakdowns caused by underlying equipment conditions. The purpose of proactive maintenance should be identifying, anticipating, and eliminating before they actually occur. Creating a proactive maintenance program characterized by industry 4.0 helps organizations find hidden inefficiencies as described by [29].

The second pillar of the technology view presented in this taxonomy is the ***technology*** employed. It should have some characteristics prone to industry 4.0 as explained below.

- *Information and communications technology* refers to respective technology used to handle telecommunications, broadcast media, intelligent building management systems, audiovisual processing and transmission systems, network-based control and monitoring functions, etc. Converging technologies that exemplify ICT should also include the merging of audiovisual, telephone, and computer networks through a common cabling system.

- *3D printing* or *additive manufacturing* is the process of making three-dimensional solid objects from a digital file. The creation of a 3D printed object which is achieved using additive processes is considered to be an important aspect of industry 4.0. In an additive process, an object is created by laying down successive layers of material until the object is formed. Adapting industry 4.0 to any enterprise should have this capability not just for producing the products but also for understanding the respective product behavior.

- A *smart sensor* is a device that takes input from the physical environment and uses built-in compute resources to perform predefined functions upon detection of specific input and then process data before passing it on. Since they operate interactively over an IT network, they are called cyber-physical systems. Industry 4.0 is heavily based on these sensors and respective Internet of Things.

- *Artificial intelligence* is, on the other hand, one of the pioneering technologies facilitated by industry 4.0, especially for ensuring required autonomy and self-behaving capability. Research associated with artificial intelligence is highly technical and specialized. Knowledge engineering is a core part of AI research. Machines can often act and react like humans only if they have abundant information relating to the world. Artificial intelligence must have access to objects, categories, properties, and relationship among those.

- *Fraud detection* technology, especially those utilizing machine learning systems, is expected to become one of the top priorities of industry 4.0 implementations. Fraud detection problems are known for being extremely imbalanced. Boosting is one technique that usually works well with suspicious data sets. It iteratively creates weak classifiers (decision trees) by weighting the instances in order to increase the respective performance.

- *Human-machine interface* (HMI) technologies enable people to interact with computers. It is important to employ right set of equipment and related technologies to have continuous operations of manufacturing systems. It is known to IT technology specialists that the archaic user interfaces can reduce productivity on the machines which are commonly among the most expensive in the facilities. That encourages the industry 4.0 specialist pay attention to this technology.

- There are various technologies that can provide real-time information processing to the devices located apart. Those devices may include mobile telephones, laptop computers, personal digital assistants, and gaming consoles. *Location detection technologies*, such as the global positioning system (GPS), are included as a standard feature in many new mobile telephones. Similar technologies have to be utilized where manufacturing units are located.

- An *IoT platform* is an indispensable technology for industry 4.0. A multilayer structure that enables straightforward provisioning, management, and automation of connected devices within the universe of Internet of Things (IoT) is set up. IoT platforms were originated in the form of IoT middleware, whose purpose was to function as a mediator between the hardware and application layers. Its primary tasks include data collection from the devices over different protocols and network topologies, remote device configuration and control, device management, and over-the-air firmware updates. There is a need to employ the platform for easy implementation of these.

- Similar to those listed above, *mobile devices* are to be used within the manufacturing suits in industry 4.0 more and more everyday. New data storage, processing, and display technologies have allowed these small devices to do nearly anything that had previously been traditionally done with larger personal computers. These devices should be configured in accordance with the purpose of the manufacturing system.

- *Augmented reality (AR)* is an interactive experience of a real-world environment whose elements are augmented by perceptual information generated by computer simulation. Integrating real and synthetic worlds is also the basic requirement of industry 4.0. Since augmented reality is the ability to integrate digital information into real-time experiences, this technology will remain at the top of the agenda in implementation projects.

- *Cloud computing* is the delivery of computing services—servers, storage, databases, networking, software, analytics, intelligence, and more—over the Internet (the cloud) to offer faster innovation, flexible resources, and economies of scale. As mentioned above, cloud computing technology is to be empowered within manufacturing and respective management activities.

In order to utilize the technologies listed above over the IT, network requires a set of information processing technologies to be embedded into the system. That is why enterprise resource planning (ERP) related to work has a remarkable impact on industry 4.0 [30–32]. In manufacturing suits equipped by industry 4.0 components, the following ***information processing systems*** need to be equipped and operated.

- *A manufacturing execution system (MES)* which is a software system connecting monitors and respective control devices to manufacturing systems and handling data flows that take place on a factory or shop floor. The overall goal of MES is to make certain that manufacturing operations are effectively executed and improve production output. This is essential when there is not much human involved in the operations as implied by industry 4.0.

- The basic software framework to support information processing could be *ERP systems*. They should be obliged to take into account innovations such as the industrial Internet of Things (IIot), which is revolutionizing manufacturing by leveraging intelligent, connected devices in factories. Better data and process analysis capability ensures even more opportunities to fine-tune their operations.

- The fourth industrial revolution and the rise of things such as IoT, robotics, and sensors could in turn create a system which can produce things much more efficiently. *Big data* and *analytics* such as machine learning methodologies can be utilized in order to dig relevant information out of the data collected. This may make predictions and spotting right knowledge easy.

- *Customer profiling* is a way to create a portrait of customers to help in making design decisions concerning service. This facilitates customizations and support customer satisfaction.

Digital manufacturing which is also an indispensable component of industry 4.0 is roughly the application of digital technologies to manufacturing. It is all about having the right information, at the right place, at the right time. The goal is to link disparate systems and span processes across all departments and functions within the value chain.

- Digital modeling and fabrication is a design and production process that combines *3-D modeling* or computing-aided design (CAD) with additive and subtractive manufacturing. It is therefore considered to be one of the essential parts of the proposed taxonomy.

- As stated earlier *IT systems* with respective sub-technologies should be utilized within manufacturing units. Mobile devices will be used as the main applications in order to manage so called remote systems. Today mobility already provides solutions that allow an employee or manager perform vital management functions from a phone or tablet, and do so from wherever they work or stay.

- Additive manufacturing provides some extra opportunities for manufacturing society. These include more inspections with fewer devices, reliable measurement, and detection of challenging target, cost savings, IO-link communication, real-time remote monitoring and diagnostics, and predictive maintenance. All of these new technologies necessitates the definition of new *quality control processes*. Traditional quality practices will not be sufficient enough to assure the required level of quality.

- *Nanotechnology* is an area that can change the whole world independently of all other criteria. It is expected to benefit greatly from the production stage in the near future. İndustry 4.0 applications should not disregard the progress on nanotechnologies.

3.4 Human resource view

There is a strong relationship between industry 4.0 and human life or society. When industry 4.0 is effective all over the world, business life, human resources, and the social lives of people are foreseen to have complete change. Although much of the operational processes are carried out by robots and machines, industry 4.0 will lead to major changes on people abilities which is highlighted in **Figure 5**. In this taxonomy, four aspects of human resources are considered to be worth mentioning as explained below.

The first component of these categorizations is the ***people*** pointing out mainly the required skill and competencies of the employees from the following points of view.

- *Employee autonomy* can have positive benefits for both the individuals and the company as a whole. When staff members have the freedom to approach their jobs in a way that suits their personalities and skill sets, business that can flourish from all aspects are covered. As planning a way to strike the right balance of autonomy in the company, it is important to define how employees can work within the structure of the business in the most effective way.

- *Employee openness*, trust, and transparency also ensure that people are kept informed of issues as they arise and that everyone receives current knowledge about an ever-changing situation. Lack of openness can create distrust, lead to hidden agendas, and be detrimental to the team, attitude, work, and productivity. Company and business owners who value the thoughts and opinions of their staff stand much better chance of being successful. In the future this openness will be more required than today.

- Employers want effective and *competent employees*, regardless of the position of their job. There are ways to evaluate prospective employees for specific types of jobs. Machine autonomy will surely be empowered by staff competency in computing, programming, behavior modeling, simulation, real-time information processing, etc.

- A *skill set* is a particular category of knowledge, abilities, and experience necessary to perform a certain type of a job. Industry 4.0 practitioners should also identify the required set of skills to perform daily operations where human is involved.

- Similarly, required skills may not be found right-hand. There may be a need for *skill acquisition* theory which is not just a theory of the development of language, rather it is a general theory of learning ranging from cognitive to psychomotor skills. This theory, which is based on, for example, adaptive

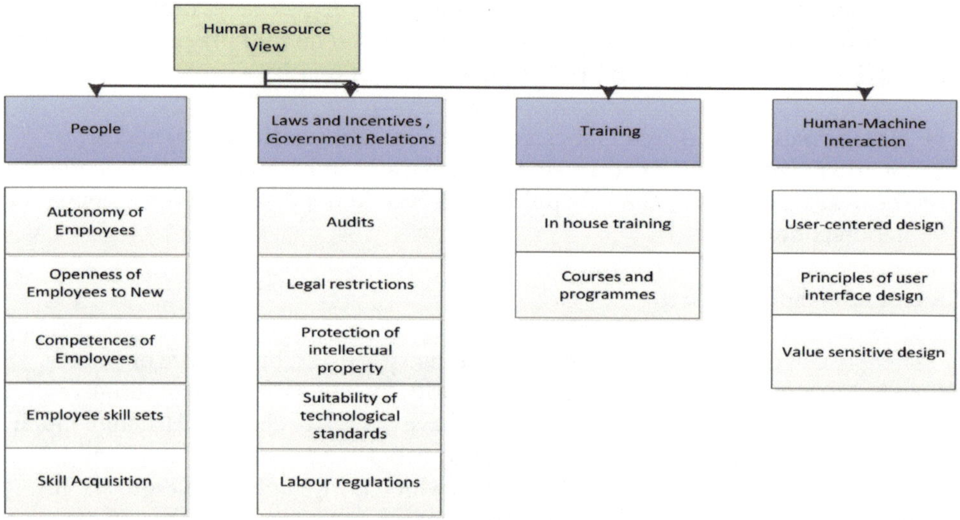

Figure 5.
Human Resource View of Industry 4.0.

control of thought model, claims that adults commence learning something through mainly explicit processes and, through subsequent sufficient practice and exposure, proceed to implicit processes.

Laws, ***legal restrictions***, ***protection rules***, ***employee rights***, and audits will change rapidly in the near future. Decentralized business processes will have to be examined in audits as opposed to centralized audits of today. "Auditing" procedures should be redefined to cope with industry 4.0 baselines. Similarly, "legal restrictions" and employee rights will be revised taking the man–machine interfaces into account. There must be a change in the perception of classical management and the possibility of working more freely. As the production speed and capacity will increase, the competition should go much more controlled. The protection of the environment, human, and society will be essential. Intellectual property rights (IPR) systems can be critical in helping new ventures transform their innovation potential and creativity into market value and competitiveness. Efforts should be continued to ensure that "technological standards" are sustainable, and technology audits should be increased while keeping respective IPR on hand. This will definitely yield "labor regulations."

It is now certain that in industry 4.0, single-level operational jobs, low-quality jobs that do not require qualification, and jobs that require physical power will be carried out by robots. Simulation and similar studies will heavily be used for staff training and maintenance. Virtual reality will make a great contribution to training in this regard. It is expected that the education and training system will totally change. New topics within the scope of industry 4.0 will have to be covered. Education system will not be enough to create the required level of competency and skills which in turn will necessitate "in-house training." On-site training is believed to increase rapidly in the near future. Technical knowledge and specialization will gain more importance than today. Programs and trainings will also be multifaceted.

While training is one of the valuable tools to support industry 4.0 in terms of mobility and access, ***man-machine interaction*** enables more flexible operation from the following aspects.

- Today, many processes are carried out by people. However, they will be replaced by machines. People already started to work with chatbots and digital assistants. Online shopping, online banking, and after-sales processes are now changing rapidly. It is important that people have *user-oriented design* features to be satisfied with their experience. Industry 4.0 will enforce designers to perform customized design as much as possible.

- Since customers will be more concerned with user-oriented design features, designing interfaces that everyone can use will then easily contribute to competition, that is to say that the user-friendly products will typically be more successful than those with complex, convoluted interfaces. Additionally, customers often avoid unreliable products, such as software programs that are full of bugs. In order to ensure a good user experience, companies often thoroughly test their products before releasing them to the public. Especially, software products will be designed obeying the principles of user interfaces attracting the attention of users to important points and issues.

- Similar to user centric design and user interface design, *value sensitive design (VSD)* will also be the point of attention in industry 4.0 implementations. VSD is a theoretically grounded approach in order to make the design of technology that accounts for human values in a principled and comprehensive manner. Accurate analysis is needed to rapidly adapt to user expectations. Interfaces must be interchangeable and renewable according to the individual's wishes. Customized processes and products of industry 4.0 will definitely require VSD to be at the top agenda item of designers.

4. Conclusion

Being able to cope with digital transformation requirements, the companies should pay attention to certain phenomena interrelated to one another. It is now obvious that future manufacturing will be more intelligent, more flexible, more adaptive, more autonomous, more unmanned, and more sensor based (industry 4.0 standards). More and more augmented reality will take place in production suits. This will naturally change the man power profile as well. There is a need to carry some research along this line. Moreover, future manufacturing systems will not only be based upon industry 4.0 standards but will be more extended toward generating fully automated and unmanned systems with having robots enriched with humanlike behaviors. The correct use of real-time information is expected to lead not only today but also the next industrial revolution.

This new trend will not only affect the economy and manufacturing industries but the whole society, education, health, and law. It can be considered a well-accepted start that will guide the human life toward the future societies. Today, all of the components considered in this research are assembled to some degree one way or another to serve economies. In order to facilitate this transformation, the taxonomy of industry 4.0 is developed and presented in this chapter. Change management process along this line could be effectively managed by concentrating and utilizing the components of the defined taxonomy which consider strategic, managerial, technological, and human interactions from various aspects.

The taxonomy presented could obviously be expanded and aligned to specific field with some sector-specific needs. The impacts of biotechnology and nanotechnology applications could be elaborated in more detail. Since industry 4.0 will have a severe effect on politics, trade, and the legal systems, there could be a need to extend the taxonomy in the light of the developments for better reflecting the reality.

On the other hand, industry 4.0 will change working life and business processes of people as well. For this reason, studies on labor economics and labor rights could be considered in terms of their effect on manufacturing and service providing systems. The taxonomy will help improve those, and the achievements in turn will help improve the taxonomy.

Author details

Ercan Oztemel* and Samet Gursev
Faculty of Engineering, Department of Industrial Engineering, Marmara University, Istanbul, Turkey

*Address all correspondence to: eoztemel@marmara.edu.tr

References

[1] Shafiq SI, Sanin C, Toro C, Szczerbicki E. Virtual engineering object (VEO): Toward experience-based design and manufacturing for industry 4.0. Cybernetics and Systems. 2015;**46**:35-50

[2] SmartFactory KL. Keyfinder production line. 2014. Available from: https://smartfactory.de/wp-content/uploads/2018/04/SF_WhitePaper_Safety_3-1_EN_XS.pdf [Accessed: 07 June 2019]

[3] Oztemel E, Gursev S. Literature review of industry 4.0 and related technologies. Journal of Intelligent Manufacturing. 2018:1-56. Available from: https://link.springer.com/article/10.1007/s10845-018-1433-8#citeas

[4] Schumacher A, Erol S, Sihna W. A maturity model for assessing industry 4.0 readiness and maturity of manufacturing enterprises. Reconfigurable and Virtual Production. 2016;**52**:161-166

[5] MetamoFAB. 2017. Available from: https://www.festo.com/group/en/cms/10275.htm [Accessed: 07 June 2019]

[6] Klaus H. On the way to industry 4.0-a Digital Enterprise, Siemens industry 4.0 report. 2015. Available from: https://www.siemens.com/press/pool/de/events/2015/digitalfactory/2015-04-hannovermesse/presentation-e.pdf [Accessed: 07 June 2019]

[7] Mckinsey. Industry 4.0: How to navigate digitization of the manufacturing sector. 2016. Available from: https://www.mckinsey.com/business-functions/operations/our-insights/industry-four-point-o-how-to-navigae-the-digitization-of-the-manufacturing-sector [Accessed: 07 June 2019]

[8] PNC. PNC industry 4.0 report. 2016. Available from: https://content.pncmc.com/live/pnc/corporate/pncideas/articles/CIB_ENT_PDF_1015-0136-198919-Factory_of_the_Future_Sept_Articles_final.pdf [Accessed: 07 June 2019]

[9] Rockwell. You cannot achieve Smart Manufacturing without embracing modern technology. 2016. Avaibale from: https://literature.rockwellautomation.com/idc/groups/literature/documents/sp/cie-sp005_-en-p.pdf [Accessed: 07 June 2019]

[10] Platform i40. What is Industry 4.0. 2016. Available from: https://www.plattform-i40.de/PI40/Navigation/EN/Industrie40/WhatIsIndustrie40/what-is-industrie40.html [Accessed: 07 June 2019]

[11] Accenture. Industry 4.0 Revolution Report. 2016. Available from: https://www.accenture.com/us-en/insight-digital-industry-impact [Accessed: 07 June 2019]

[12] Kagermann H, Wahlster W. and Helbig J. Recommendations for implementing the strategic initiative Industry 4.0: Final report of the Industry 4.0 Working Group. 2013. Available from: https://www.din.de/blob/76902/e8cac883f42bf28536e7e8165993f1fd/recommendations-for-implementing-industry-4-0-data.pdf [Accessed: 07 June 2019]

[13] Cooper S. Rethink Manufacturing, Designing a UK Industrial Strategy for the Age of Industry 4.0. Technical Report, KPMG. February 2017

[14] Rosendahl R, Schmidt N, Lüder A, Ryashentseva D. Industry 4.0 value networks in legacy systems. In: IEEE 20th Conference on Emerging Technologies & Factory Automation (ETFA), 8-11 Sept. 2015, Luxembourg. 2016. pp. 1-4

[15] Bourke R, Mentis M. An assessment framework for inclusive education: Integrating assessment approaches. Assessment in Education. 2014;**21**(4):384-397

[16] Suh Y, Lee H. Developing ecological index for identify roles of ICT industry in mobile ecosystems. Telematics and Informatics. 2017;**34**(1):425-437

[17] Qin J, Liu Y, Grosvenora R. A categorical framework of manufacturing for industry 4.0 and beyond. Virtual Production. 2016;**52**(2016):173-178

[18] Adeyeri S, Kanisuru M, Khumbulani M, Olukorede T. Integration of agent technology into manufacturing Enterprise: A review and platform for industry 4.0. In: Proceedings of the 2015 International Conference on Industrial Engineering and Operations Management, Dubai, United Arab Emirates (UAE). 2015. pp. 1625-1635

[19] Schouh G, Gartzen T, Marks A. Promoting work-based learning through industry 4.0. CIRP Conference on Learning Factories. 2015;**32**:82-87

[20] Brandmeier M, Bognera E, Brossoga M, Frankea J. Product design improvement through knowledge feedback of cyber-physical systems. Procedia CIRP. 2016;**50**:186-191

[21] Jeang A. Robust product design and process planning in using process capability analysis. Intelligent Manufacturing Systems. 2015;**26**(3):459-470

[22] Bello O, Zeadally S, Badra M. Network layer inter-operation of device-to-device communication technologies in internet of things (IoT). Ad Hoc Networks. 2017;**57**:52-62

[23] Gudfinnsson K, Strand M, Berntsson M. Analyzing business intelligence maturity. Journal of Decision Systems. 2015;**24**(1):37-54. DOI: 10.1080/12460125.2015.994287

[24] Hubert C, Chan Y. Internet of things business models. Journal of Service Science and Management. 2015;**50**:1020-1030

[25] Jourdan Z, Rainer K, Marshall T. Business intelligence: An analysis of the literature. Information Systems Management. 2008;**25**(2):121-131

[26] Strozzi F, Colicchia C, Creaazza A, Noe C. Literature review on the smart factory concept using bibliometric tools. International Journal of Production Research. 2017;**55**:1-20. DOI: 10.1080/00207543.2017.1326643

[27] Wang S, Wan S, Zhang D, Li D, Zhang C. Towards smart factory for industry 4.0: A self-organized multi-agent system with big data based feedback and coordination. Computer Networks. 2015;**101**:158-168

[28] Theorin A, Bengtsson K, Provost J, Lieder M, Johnsson C, Lundholm T, et al. An event-driven manufacturing information system architecture for industry 4.0. International Journal of Production Research. 2016;**55**:1297-1311

[29] Sipsas k, Alexopoulos k, Xanthakis V, Chryssolouris G. Collaborative maintenance in flow-line manufacturing environments: An industry 4.0 approach. Research and Innovation for Future Production. 2016;**55**:236-241

[30] Tong Li YW, Junjiaou W. Capacity analysis of an iron foundry fettling-shop, using virtual manufacturing technology. International Journal of Cast Metals Research. 2016;**16**:329-332

[31] Yang X, Malak R, Lauer C, Weidig C, Hagen H, Hamann B, et al. Manufacturing system design with

virtual factory tools. International Journal of Computer Integrated Manufacturing. 2013;**28**(1):25-40

[32] Seethamraju R, Sundar D. Influence of ERP systems on business process agility. Management Review. 2013;**25**(3):137-149

Section 2

Case Studies

Chapter 5

AUTO 4.0: Anticipation of Skills for Employees Due to Digitalization - Identification of "Occupational Profiles"

Georg Spoettl

Abstract

Europeanization is a politically supported process which includes mobility of the labor force, high-service quality, fast use of modern technology, and a chance of interaction of companies and other stakeholders. The economic activities of companies are worldwide and global—the world as a global village. The new world is change-driven, and radical developments for businesses based on the use of high-technology and data are supported by countries and companies. It is expected that with the help of Industry 4.0 the business processes will be more efficient and productive. "Globalization is the thread that ties nations together, with innovation around technology imperatively affecting trading activities" (Kalio). One of the main questions is about the impact of training on the development process of Industry 4.0 and all its implications. This question is in the focus of this paper dealing with Industry 4.0 in the automobile industry in European countries.

Keywords: AUTO 4.0, skills, occupational profiles, digitalization, industry 4.0

1. Introduction

Europeanization is a politically supported process which includes mobility of the labor force, high-service quality, fast use of modern technology, and a chance of interaction of companies and other stakeholders. The economic activities of companies are worldwide and global—the world as a global village. The new world is change-driven, and radical developments for businesses based on the use of high technology and data are supported by countries and companies. It is expected that with the help of Industry 4.0[1] the business processes will be more efficient and productive. "Globalization is the thread that ties nations together, with innovation around technology imperatively affecting trading activities" ([1], 167). One of the main questions is about the impact of training on the development process of

[1] Industry 4.0 is the current trend of automation and data exchange in manufacturing technologies. It includes cyber-physical systems (CPS), the Internet of things (IoT), and cloud computing. The term is mainly used for manufacturing processes in different sectors. For the description of the overall changes in societies, the term "4th Industrial Revolution" is in use, and for the concrete changes in different business fields, the term "digital transformation" is applied.

Industry 4.0 and all its implications. This question is in the focus of this paper dealing with Industry 4.0 in the automobile industry in European countries.

2. Dimensions of change through digitalization

There still is an ongoing discussion of the impact of digitalization[2]. A very common statement: "For the majority of workers and employees, the immediate effects of Digitalization are probably not visible yet" ([2], 86). Besides there are convincing indicators regarding the progress of digitalization with the economy (cf. ibid). More and more companies are running their business with the help of digitalization, among them, Microsoft, Apple, Facebook, Uber, and others. Even small and medium-sized companies are increasingly relying on digital technology. An example: In the city of Hamburg, around 92,000 companies are registered. Ninety percent of these companies apply digital equipment to optimize the business and work processes (cf. [3], 14). An empirical study shows this movement. In order to assess the diffusion depth of "Industry 4.0" and thus its presence in companies of the metal and electrical industry, including the automotive industry, the authors of the bayme vbm Study ([4], p. 56) have developed an instrument for the assessment of the diffusion of technology and work organization. The former encompasses seven technology dimensions such as sensor technology/actuating elements (networking CPS), networking (entire value-added chain), radio technology (communication), Big Data (data analysis), cloud computing (date storage, data speed), workplace intelligence CPS (share of man and technology) up to data safety (data sovereignty). Within expert workshops, each of these dimensions was assessed by experts with a view to the diffusion depth of "Industry 4.0" in companies. The result (cf. **Figure 1**) is the midpoint value of all experts. Thus a reference system was created which allows for a clear-cut characterization of the development steps towards "Industry 4.0" related to the dimensions of technology. The result presented in **Figure 1** indicates the implementation depth of "Industry 4.0's" technology as assessed by the experts.

Figure 1 also indicates that sensor technology and actuating elements have reached the highest markedness. This means that the communication of the CPS via

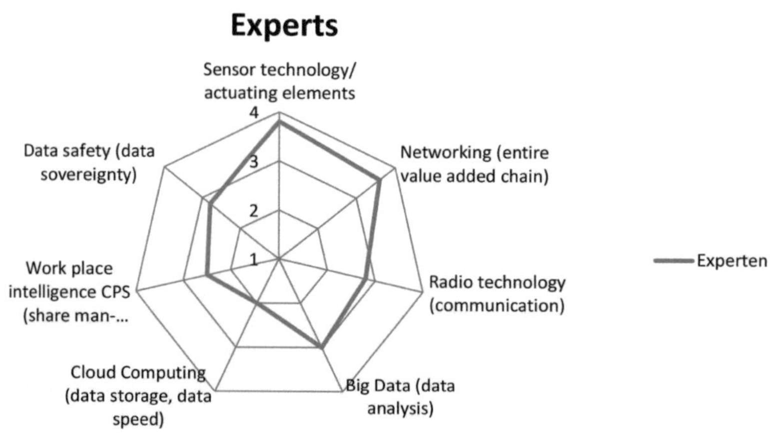

Figure 1.
Diffusion steps of technologies—Assessment by experts (source: [5]).

2 The term can be understood as the technological issue of Industry 4.0.

digital communication devices has already been networked within the value-added chain and that this is highly relevant for the level of skilled work. A minor rating—markedness level 2—was assigned to the dimension of cloud computing. Data storage has so far mostly been taken care of by the companies themselves and still has a considerable potential for development. Both radio technology and workplace intelligence were also low-rated. The reason for the low rating of workplace intelligence could lie in the fact that it is just being implemented at an early stage. Data safety is an issue linked to a lot of trust which can apparently not yet be guaranteed (cf. [5]).

The target of digitalization/Industry 4.0[3] is to make business processes more efficient and productive. However, an adoption of Industry 4.0 will result in rapid change of the job tasks triggered by automation. There will be a deep impact on lower-skilled jobs. This transformation will require a significant change of the workforce's skills, organizational structures, leadership mechanisms, and corporate culture ([6], 137).

According to Schwab [7], ubiquitous mobile supercomputing, intelligent robots, self-driving cars, neuro-technological brain enhancement, and genetic editing are evidence for a global dramatic change taking place at exponential speed (cf. [1], 169). Schwab's statement: "We are at the beginning of a revolution that is fundamentally changing the way we live, work and relate to one another" [7]. He expects that all disciplines in economies and industry and the quality of life will be confronted with these changes.

3. Impact of industry 4.0/digitalization on vocational education

Looking at industry 4.0 and globalization, it is hard to say that organizations are ready to adapt to the big change of the future, knowing that they bag a heavy load of challenges with respect to technological, social, and ethical advancement.

Marr [8] emphasized the need to upgrade skills at the workplace and we must learn to understand and collaborate with the intelligence of machines at our disposal.

The changes in the world of work have a deep impact on the training needs. Three categories are the core of these changes:

1. Technology

2. Work organization

3. Social and ethical dimensions

They form the transformation of the workplace. An organization of the future is one in which employees will no longer be able to rely on simple skill training or facts of discipline, but development as an integral part of the job will necessitate on-the-job-training and retraining up with the demands (cf. [1], 170). In this context work-based learning and the use of virtual technology, simulation, digital media, and others become more and more important. Learning from the future is taking a different dimension. The integration of learning, work requirements, virtual

---

[3] In the article the term "Industry 4.0" will be used with high priority because of the production sector of the automobile industry as a subject of survey.

technology, and different methods of learning will become more important. Workplaces require highly skilled workers for a broader[4] and technology-driven organization:

"Industry 4.0 can rightfully be referred to as a production paradigm, since we have on the one hand intelligent factories and on the other hand production and logistics processes which are globally interconnected over the internet. This enables a flow of materials which can be optimized and interconnected to a degree so far unknown.

Due to digitalization and enrichment with information, work- and business processes, web-based and mobile as well as services based on intelligent analyses of large data bases are becoming more and more important and they achieve a remarkable impact on the design of high-tech work environments and hence on the workplaces involved.

This kind of technological development has definitely to be addressed as a long-term strategical project, which intends to create intelligent closed processes in production, the neighboring fields as well as finally within the entire value-added chain of production. This calls for innovative concepts of interaction between man and machines in order to direct work processes in the future [9].

Industry 4.0 is creating a vacuum of knowledge and capabilities for the employees, especially for those within the challenging environment in companies with a higher diffusion of digitalization technologies.

In order to find convincing answers for the qualification of the workforce on shop floor level in the context of Industry 4.0, it is crucial to clarify the need for qualification for the skilled workforce in the companies.

4. Identification of "occupational profiles" for industry 4.0

Vocational education and training (VET) has high priority due to the manifold requirements of industry. The VET priorities of the European Commission (cf. [10]) are highly relevant because of Industry 4.0:

- The aim is to provide workers with continuing training programs for upgrading of their skills and in order to respond quickly to emerging needs.

The new guidelines (cf. [11]-381) require:

- To build resilience through the development of key skills and higher and complex skills
- To focus on the skills needed by full and complete digitalization of industry and services
- To strengthen enterprises, VET providers, and partnerships

In Project AUTO[5] these statements were in the center of activities.

[4] This includes social and ethical dimensions.

[5] It is an ERASMUS Project with the title: AUTO 4.0—Understanding and Achieving Automotive Training Outcomes 4.0 - *Erasmus + Programme 2017-1-IT01-KA202-006187.*

4.1 The target group and requirements

One of the core objectives in the first period of the project AUTO 4.0 (Understanding and Achieving Automotive Training Outcomes 4.0) with partners from Italy, Spain, Great Britain, and Germany was the identification of "qualification profiles" and/or "occupational profiles" for workforce in the automobile production. This objective was linked to the question of the kind of methodological approaches that should be applied in order to achieve insights and results. With regard to the participating partnership, this question was not marginal. The situation of the partners and thus the access to the automobile industry—the topic the project concentrated on—were as follows:

- Partners from Italy: Access above all to component manufacturers
- Partners from Spain: Access to car manufacturers and component manufacturers
- Partners from Great Britain: Access to a network of manufacturers in the automobile sector
- Partners from Germany: Access to car manufacturers and component manufacturers

In order to answer the central questions of the project, the project consortium agreed on a common empirical approach (see below). The target was to identify occupational profiles (or other solutions) as an answer to the changes within the fourth industrial revolution. It was defined that the holder of competences – based on occupations or other solutions – are able to master the relevant tasks of his/her field of activities, that he/she can contribute to a high quality of the product and that he/she is especially creative. Thus it contradicts the general opinion of occupations and professionals of the trade. In order to safeguard this development, it is necessary to keep ordinances—i.e. the control mechanisms for a vocational education—up to date.

4.2 Method for the determination of qualification profiles

In order to ensure that all partners concentrated their activities on the identification of skills and occupational profiles of comparable fields, it was agreed that the surveys of component manufacturers (preference on first tier suppliers—the so-called system suppliers) and car manufacturers should be given priority. Within the survey, instruments were applied—such as case studies in well-selected companies and expert discussions—to ensure a deep insight into the changes in companies due to the use of Industry 4.0-driven technology.

Two categories of guiding questions were developed for the surveys. Guiding questions were necessary because the surveys concentrated on expert conversations. On the other hand, the guiding questions were important for the conversations with different target groups in the companies.

The guiding questions for expert discussions were aiming at managers and experts of Industry 4.0.

In each partner country, five managers and five experts were interviewed. The discussions were strongly focused on the impact of Industry 4.0 on the companies and on how the employees were qualified for these new challenges. The results of these interviews were applied to identify the need for qualification.

As for the case studies, separate guiding questions were developed to interview the following target groups in companies:

- Human Resources Directors (HR Directors and experts of Industry 4.0 implementation)
- Recruitment experts
- Skilled workers
- Experts for training.

An example of the guiding questions for experts and HR Directors is attached in Annex I.

Each partner country planned case studies in the following kinds of companies:

- Car manufacturers (1 case)
- System suppliers (first tier) (1 case)
- Suppliers on a lower level (1 case)

With the aid of the case studies, the developments in the selected companies could be thoroughly opened up. The focus was above all on the technological and work-organizational changes that have taken place in the companies. In addition, the surveys revealed the measures taken by the companies to qualify their staff for the new requirements. All partners conducted expert interviews (in total 25). As for the case studies, however, some gaps remained.

The findings of these surveys formed the basis for the design of qualificational profiles.

4.3 Results of the survey phase

4.3.1 Identification of competence profiles

A form which was used by all partners for the documentation of the identified competence profiles was developed in order to facilitate the agreed documentation process and the results of the empirical surveys as requirement profiles across all countries. Apart from this, it was possible to develop supplementing documents.

The "qualification" or "competence profiles" developed with the uniform format helped to document the requirements in employment fields which are intensively permeated by Industry 4.0 and call for competences so far not necessary for employees.

Competences were generated from the requirements for the employees and formed the basis for the description of competence profiles and/or occupational profiles.

Thus a total of 19 occupational profiles were identified which are playing a role in the surveyed companies and which are initiated by the development of Industry 4.0.

The topics of the 19 profiles are concisely shown in **Figure 2**.

The following profiles were developed in the individual partner countries:

	Qualification Profiles	Countries			
		ES	DE	IT	UK
1	Design Project Leader	(L5/6)			
2	Maintenance Team Leader	(L4/5)			
3	Quality Technician	(L5)			
4	Production Supervisor / Shift Supervisor	(L5/6)			
5	Mechatronic Expert	(L4)			
1	Data an Process Management		(L5/6)		
2	IT Systems and Networks		(L5/6)		
3	Trouble Shooting and Problem Solving		X (L5)		
4	Maintenance and Repair		(L5)		
1	Mechatronic Operator 4.0 (Operating Technician in Automotive ...)			(L5)	
2	Supplier Quality Assurance 4.0			(L6)	
3	Technologist New Production Processes			(L5/6)	
4	Cyber Security Technician			(L6/7)	
5	Data Scientist			(L6/7)	
6	Data Architect			(L6/7)	
1	Design Engineer				(L6/7)
2	Project Engineer				(L6/7)
3	Operator Manufacturing				(L4/5)
4	Senior Technician				(L5)

Figure 2.
"Qualification": Profiles of "industry 4.0" for the European automotive industry. ES = Spain, DE = Germany, IT = Italy, UK = United Kingdom.

1. **Spain:**

   Design Project Leader

   Maintenance Team Leader

   Quality Technician

   Production Supervisor/Shift Supervisor

   Mechatronic Expert

2. **Germany:**

   Data and Processes Management

   IT Systems and Networks

   Troubleshooting and Problem Solving

   Maintenance and Repair

3. **Italy:**

   Mechatronic Operator 4.0 (Operating Technician in Automotive)

   Supplier Quality Assurance 4.0

   Technologist New Production Processes

   Cyber Security Technician

   Data Scientist

   Data Architect

4. **Great Britain:**

   Design Engineer

   Project Engineer

   Project Engineer

   Operator Manufacturing

   Senior Technician.

Based on the description of competences, it is possible to allocate the individual profiles to the levels of the European Qualifications Framework (EQF) (see **Figure 2**). This was done by an expert assessment of project members rather than by a systematical comparison of individual indicators. Example: (L 5/6) means that the quality of the profiles is between level 5 and 6 of the European Qualification Profile (EQF).

4.3.2 Requirement level of competence profiles

The result of the allocation of all of the profiles is shown in more detail in **Figure 2**. It is obvious that mainly levels 4, 5, 6, and 7 are applied. Thus the profiles not only cover the requirements that are playing a role in vocational initial training and further training. In addition, also academic profiles come into effect. Thus the profiles cover a very comprehensive spectrum. It is remarkable that profiles on levels 4 only play a minor role and profiles on level 3 are mentioned once only.

Consequently it can be said that cognitive profiles and theoretical and more demanding profiles are dominating. **Figure 3** underpins this general assessment in more detail. When asking about the role of cognitive challenges during data processing and in work processes, the majority of the named competences underline a very high level (Quadrant I of the Matrix). Fourteen profiles of this criterion

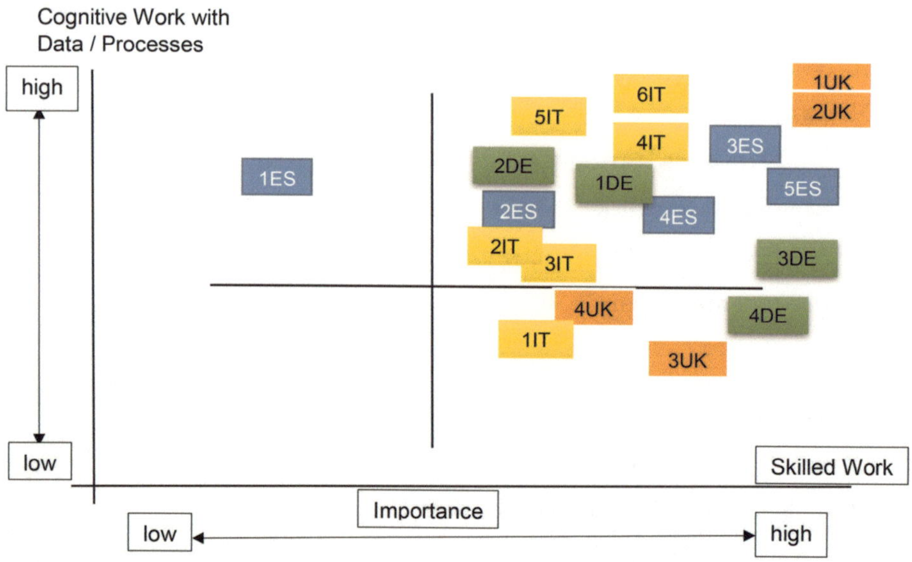

Figure 3.
"Competence quality" of the occupational profiles.

have to be allocated to Quadrant I. As for "skilled work," the estimation of the quality of competences is very high (Quadrant I as well). This means that the workers are not only performing simple mechanical skills but must above all master processes and the handling of data.

Only four profiles indicate less demanding, but still high cognitive requirements and are therefore allocated to Quadrant IV. However, the skills are still comparatively demanding in these four profiles. One of the profiles shows considerably less requirements for skills but comparatively high cognitive demands. This profile was allocated to Quadrant II.

As a summary it can be noted that all profiles are influenced by the:

- Increasing implementation of Industry 4.0
- Intensification of the work processes
- Increasing networking of data

This leads eventually to very demanding competence profiles, starting with EQF level 5 and higher.

Figure 4 documents the outcome of a case study of a German supplier. It is remarkable that a clear differentiation between requirements of technology, work, and social implications is possible. It is a demonstration that technology is not the only driver of the implementation process of Industry 4.0. For a successful implementation, the structure of work organization and the requirements of work are

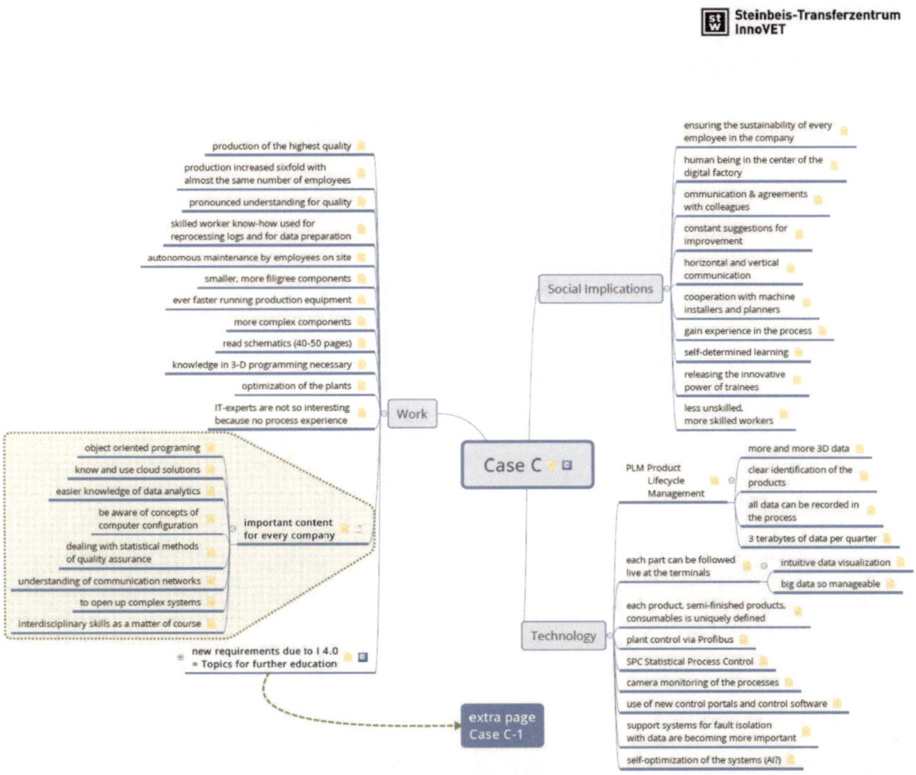

Figure 4.
Example of occupational requirements (source: Case study, Germany).

important. The shaping of the work organization decides which level and differentiation of the qualification are needed. It is also important to consider the social impact and to prepare the workforce for these tasks. Communication, cooperation, innovation, higher cognitive skills, acquiring experience in operating complex technology, continuous quality improvement, etc. are some of the social implications the workforce has to be trained for. Another issue is dealing with different kinds of software. This task requires workers who are able to make use of software, who take care of minor program modifications, or who are analyzing statistical messages which are transported via the software. Therefore workers have to be trained to enable them to shape their work in a successful way.

5. European core profile

Another question in the project was whether it would be adequate to generate a European Core Profile on Levels 4 or 5 of the European Qualifications Framework out of the numerous individual profiles. The analysis of the individual profiles shows that based on the "operation area", five individual profiles can be identified that reveal high affinity in terms of competence requirements and in terms of contents priorities.

Figure 5 gives an overview. The "operation area" consists of the following profiles:

- Mechatronic expert
- Maintenance team leader
- Mechatronic operator 4.0
- Maintenance and repair

The profiles are all linked to levels 4 and 5 of the EQF-Levels. The quality of the profiles is comparable, and the contents of work do not differ much. All the profiles have the aim to ensure the operation of plants.

The "operation area" is the level where plants are taken in and kept in operation by exclusively employing skilled workers who have undergone training and further training below the academic levels and who are working on the shop floor.

Core Profiles	ES	DE	IT	UK
Operation Area				
Mechatronic Expert	L 4			
Maintenance Team Leader	L 4/5			
Operator Manufacturing				L 4/5
Mechatronic Operator 4.0			L5	
Maintenance & Repair		L 5		
Quality Area				
Quality Technician	L 4/5			
Supplier Quality Assurance			L 6	
Data Area				
Data Architect		L 6/7		
Data Scientist			L 6/7	

Figure 5.
Core profiles for generating a European profile.

With the aid of a contrastive analysis, the similarities of the profiles were worked out which form the core of a European profile. **Figure 5** shows the five profiles of the "operation area" which were used by the contrastive analysis to generate the European Core Occupational Profile.

The profiles of the "quality area" and the "data area" are new profiles which differ in their quality level, contents of work, and linkage to the work organization of companies. These profiles have a new character to support Industry 4.0 in different fields and should stand alone.

After the contrastive analysis of the profiles of the operation, a detailed description of the competence profile was drafted. The result is documented in Annex 2. The project consortium has decided to name this profile **Automotive Digital Mechatronic X.0.**

This European Core Profile excels by the following innovative elements:

- Access to interconnected new technology via software

The "new basics": broad competences	
Learn to **think starting from the software** Get to know **network structures** Learn how to master **Big-Data technologies** Learn how to work with a **variety of data formats** Understand and **master processes**	Learning how to take over **more self-responsibility** Learning how to **cooperate and communicate better** Learning how to **initiate innovations** Understand and consider the **environmental and social impact** of technology choices and innovations Make use of **innovation** potential! Support **shaping competence**!
Context-specific competences	
In addition to their experience in plant operation, however, it is important that such persons develop further in the following areas of competence: • Problem solving • Understanding of integrated systems and their interconnections (from the own company but also from external systems) • Linking of different system controls • Think and work across disciplines • Getting involved in new tasks • To master processes • Application of IT technology as a tool • Necessity to think through the processes	In addition to their experience in plant operation, however, it is important that such persons develop further in the following areas of competence: • Use of the cloud and integration of various machine data/manufacturer data • "Third hand" will gain importance in the industrial context (e.g., lightweight robots) • Maintenance, monitoring, and care of drive technology • Consideration of the entire value chain • Data as raw material; use it and attach more importance to it • To work in the delimitation of space and time • Digitization must be designed • Target perspective: Mastering multifunctional plant operation
"Abstract" competences	
Creativity Creation Critical thinking Communication Collaboration (in teams) Modeling skills Data gathering and mining Respect of procedures Relational communication skills	Investigative character Analytical spirit Storytelling skills Lateral thinking Curiosity Leadership Innovative management Vision and communication Understand business problems

Table 1.
Different types of competences.

- Use of augmented reality (visual component)
- Predictive analysis in real time
- Virtual diagnosis
- Management of manifold formats of data
- Data protection (plus laws!)
- Sustainability and productivity
- Understanding of the whole work process of a company

Based on empirical work of the project Auto 4.0, the following competences were generated:

- Broad competences (as "new basics")
- Context-specific competences I and II
- "Abstract" competences

These competences are listed in **Table 1**. Based on these competences, the project group has generic competences that were generated (**Table 2**) which form the basis for the development of learning assignments for the European Core Profile "Automotive Digital Mechatronic X.04."

• Understanding of integrated systems and their interconnections (from the own company, but also from external systems), • Linking of different system controls. **1 Where?**	• Necessity to think through the processes, • To master processes, • 'Third hand' will gain importance in the industrial context (e. g. lightweight robots), • Consideration of the entire value chain, • Target perspective: mastering multifunctional plant operation. **2 Why**
• Application of IT technology as a tool, • Use of the cloud, integration of various machine data / manufacturer data, • Maintenance, monitoring, care of drive technology, • Data as raw material' – use of it and attach more importance to it. **3 How?**	• Problem solving, • Getting involved in new tasks, • Think and work across disciplines, • To work in the delimitation of space and time, • Digitisation must be designed. **4 Who?**

Table 2.
Generic competences.

6. Conclusions

The empirical work of the project partners facilitated the development of a number of qualification profiles for different levels and core points in the field of Industry 4.0 in selected companies. In order to get a transnational core profile above all for Level 5, the core profile "Automotive Digital Mechatronic X.0" was generated from the country- and company-specific profiles.

The innovative character of this profile could be underpinned, and the relevant competences were identified. This profile forms the basis for learning units and learning scenarios.

ANNEX I: Interview Guidelines for Survey AUTO 4.0

Guiding Questions for Expert Conversations in Companies (HR Directors, Industry 4.0 Experts)

Short Version

A) General questions on the company and personal questions

Company
• Products
• Branch
Person
• Field of tasks
• Function in the company
• Occupational background
Employment structure in production (changes)
• Number of employees
• Employee structure (shares & changes)
  o Skilled workers
  o Semi-skilled workers
  o Engineers (process engineers, test engineers),
  o ...

B) Experience and the technological development in the company
• State of the introduction of „Industry 4.0 "in the company? • What are the most important drivers for the application of principles of Industry 4.0? • How have processes been changing during the past years (ICT, networking, automation)? • Which technologies (for data/ information transfer, networking or the optimization of work processes, automation) have already been implemented in your company? • Which ones? Where? In which areas? What is the objective?
C) Impact on the business and work process

• Which were your tasks during the implementation process? What did you do? How? With which tools? Were external IT-services involved? Have the stakeholders (skilled workers) been involved in the process?
• How do you involve your skilled workers in implementation processes? Which competence profiles must these specialists have? Which qualifications must they have?
• What are the tasks of skilled workers or engineers? Have roles changed in this area?
• Is your company vertically integrated from the perspective of:
  o Sales (data-driven demand prediction)
  o Product (data-driven design to value, integrated product development ...)

D) Qualification requirements and support
• Which new requirements for employees and tasks do you expect? • Which employment level will be predominantly affected by changes? Will some of the levels vanish? • Which most important competences should be imparted in future training? Are there differences compared to today? • How do typical qualification paths in companies look like? • Which additional qualifications are necessary?

ANNEX II: European Core occupational profile.

European Core Occupational Profile Auto 4.0: "Automotive Digital Mechatronic Tabstopp X.0".

Name of the European Core Occupational Profile:

Automotive Digital Mechatronic X.0.

Overall description.

Detailing[6].

What does he/she do?

The tasks for qualified workers at the level of *Automotive Digital Mechatronic X.0* encompass maintenance tasks in order to safeguard a flawless plant operation. This means that they must have access to the function of plants and must be able to cope with (metal-technological, electrical, IT-based) maintenance tasks and diagnostic in plants composed of different technological systems. Difficult repair tasks are exempted and are task of the trouble shooters. The maintenance tasks also encompass preventive maintenance prepared by recording, processing, analyzing of data and visualization of operational and production data and should be ready to be applied at the production work places at all times. In addition qualified workers must also master virtually organized maintenance tasks and must apply assistance systems for trouble shooting, documentation and knowledge transfer. Has extensive knowledge of the production processes and may be able to set up/ programmes and carry out diagnostics.

Manufacturing is understood as: Produces goods and parts from raw materials by using different production processes. This may include some setting up of machinery and basic programming.

Responsibility

Qualified workers have to have the capability to use the industry-specific software products of production planning and preventive maintenance (PPS, ERP, ...) to handle the production work process at the workplace. This includes to perform the maintenance of the autonomous systems with a remote monitoring. A further requirement is to carry out maintenance of the production control systems based on Big Data and clouds with the help of diagnostic instruments.

To perform individualized maintenance of components of machines and plants by using continuous processes of data acquisition and to implement visualization software.

A prerequisite of all maintenance work is to evaluate the information on the wear and tear of plant parts from a continuous monitoring of the machines by sensors.

Working to strict safety and quality requirements, help to achieve daily production requirements in terms of quality and quantity to meet customer expectations and requirements, optimizing efficiency and maintaining operational excellence. Able to work on many/all areas of the production process with little supervision.

Results

The "Experts Automotive Digital Mechatronic X.0" produces the following results: he/she

- Quality, Cost and Delivery Metrics.
- Evaluates the information on the wear and tear of plant parts from a continuous monitoring of the machines by sensors.
- Handles the production work process by use of the industry-specific software products of production planning and preventive maintenance.
- Carries out preventive maintenance prepared by recording, processing, and visualization of operational and production data.
- Executes maintenance of the production control systems based on Big Data and clouds und carries through diagnostic processes.

Value

Maintenance and repair tasks in networked plants and individual machines are as a rule carried out by maintenance teams. The qualified workers of the teams are specialized in tasks including IT-tasks such as network analyses or IT-guided trouble shooting. They also master procedures for the identification of malfunctions (data analysis), causes for malfunction and their repair in complex, networked plants. Target perspective: Considering repair interdependencies due to networking and IT-integration of machines and plants; software updates. They are also fit for all traditional tasks in ensuring the function of the machinery.

---

[6] The long version is available via the author.

Author details

Georg Spoettl
University of Bremen, Bremen, Germany

*Address all correspondence to: spoettl@uni-bremen.de

References

[1] Kalio N. The impact of globalization and industry 4.0 on training and retraining in developing and underdeveloped nations. European Journal of Business and Management. 2019;**11**(3):167-172

[2] Harteis C. Supporting learning at work in an era of digitalisation of work. In: Bahl A, Dietzen A, editors. Work-Based Learning as a Pathway to Competence-Based Education. A UNEVOC Network Contribution. Bonn: BIBB; 2019. pp. 85-8s

[3] Hamburg. Unser Leben mit der Digitalisierung. Google: Hamburg; 2015

[4] Bayme vbm-Spoettl G, Gorldt C, Windelband L, Grantz T and Richter T. Industrie 4.0 – Auswirkungen auf Aus- und Weiterbildung in der M+E Industrie,. Studie herausgegeben von bayme vbm, Die bayerischen Metall- und Elektro-Arbeitgeber, München. Universität Bremen: Bremen; 2016 Available at: www.baymevbm.de/industrie4.0 [Accessed: 27 May 2016]

[5] Spöttl G. Skilled workers—Are they the losers of "industry 4.0"? In: Schlick C et al., editors. Advances in Ergonomic Design of Systems. Products and Processes—Proceedings of the Annual Meeting of GfA 2016. Selected Paper. Germany: Springer; 2017. p. 16

[6] Cevik SO, Ustundag A, Kadaifci Ç, Oztaysi B. The changing role of engineering education in industry 4.0 era. In: Ustungag A, Cevikan E, editors. Industry 4.0: Managing the Digital Transformation. Switzerland: Springer; 2018. pp. 137-151

[7] Schwab K. The Fourth Industrial Revolution. UK, USA: Penguin; 2016

[8] Marr B. Why everyone must get ready for the 4th industrial revolution. Forbes. 2014 Retrieved 14-02-2018. https://www.bernardmarr.com/default.asp?contentID=966

[9] Spöttl G. Development of "Industry 4.0"! – Are Skilled Workers and Semi-Engineers the Losers? 2017, 7th World Engineering Education Forum (WEEF), IEEE Xplore: 20 September 2018. Kuala Lumpur, 2018. pp. 934-951. DOI: 10.1109/WEEF.2017.8467033

[10] European Commission. Opinion on the Future of Vocational Education and Training Post 2020. Advisory Committee on Vocational Training; 3 December 2018; Brussels: 2018

[11] COM. 381 Communication from the Commission to the European Parliament. The Council, The European Economic And Social Committee and the Committee of the Regions. A New Skills Agenda for Europe. Working Together to Strengthen Human Capital. Employability and Competitiveness 381 Final. Brussels. 2016

Chapter 6

Stabilization of Digitized Processes

Felicita Chromjakova

Abstract

Stabile production processes are oriented on flexible production process; they enable implementation of digitized concepts in real production technologies. The presented chapter describes research results oriented on actual state mapping in selected industrial companies, with core orientation of people engagement in core processes connected with production digitization. Quantitative and qualitative research by selected industrial companies was realized in combination with hypothesis testing and verification. Core research problem of the presented chapter is the compatibility between employee and cobot by workplace from the ethical point of view. Based on analyze of selected workplaces in industrial companies were identified crucial decisive parameters of effective cooperation between human and cobot with subsequent formulation of conclusions for setting the rules of ethical mutual cooperation "human-robot" by production workplace. Radical innovations in industrial companies connected with the integration of cobot technologies in the production processes are strongly connected with the aim, to give the stability in the mutual connection between employee and cobot. A model of standardized co-operation between human and cobot will be presented; it is based on the key elements of the standardization process. This is a new point of view on the ethics of the workplace, where the "human-machine" cooperation replaces the cooperation model "human-cobot."

Keywords: process, stability, Industry 4.0, performance, standardization, efficiency, human, cobot

1. Introduction

Radical innovation of industrial companies, connected with the integration of human-cobot technologies are strong connected with the requirement to give the stability in the mutual cooperation. Cobots are playing an important role in the process performance, productivity and efficiency of production process. A lot of industrial companies' architects are oriented on the ethical workplace conditions, the vision is to develop the "right" cooperation climate by production human-cobot organized workplace. Already in 1921, the Czech author Karel Čapek mentioned the term "robot" for the first time. His vision was to replace complicated human work with technology that can perform well-chosen work performance while replacing a person with equal work performance. The first consideration was frustration of a worker who began to compare his work performance with the cobot performance. Second, an important vision, it was important to begin use the cobots for dangerous production jobs or for routine production workplaces with extreme physical or psychological stress on the person and body at work—something in the sense of effective and flexible replacement of human by production workplace. Human should cooperate with cobot (or intelligent machine) only by exactly defined technology interconnections. Basic assumption of this interconnection is standardized human intelligence and digital technology intelligence.

In modern industrial company's we face daily the problem of responsibility for the implementation of standardized workplace performance. The crucial research question is, how to deal the ethical aspect of production process stability and responsibility for the required process output. Is the responsible person responsible for the output assigned to perform the performance, or it is the cobot, who is responsible for the output assigned?

The sensitivity on productivity and quality of work performance is therefore on of the cornerstones of stability of digitized processes. An important phenomenon in this context is how we look at the complexly production process productivity and the process stability of human and cobot.

Crucial questions of today's production processes in industrial companies can be summarized as follows:

- how can we analyze the production workplace according the standardization of mutual cooperation between human and cobot?

- are we able to identify core pillars of effective and ethical cooperation by workplace given as a combination of human and cobot?

- it is possible to design the optimal standard of production workplace according the ethical human-cobot cooperation?

- can we apply the same rules to assessing compliance of job or workplace standards for human and cobot?

- which type of competencies and responsibilities are identified by human and cobot in the stabile digitized processes?

2. Theoretical background of digitized process stability

An important part of today's standardization procedures in industrial companies is the accent on the optimal human—cobot interaction, controllable sensors by human and cobots, flexible jobs layout organization between human and cobot and at the end the requirement on the optimal level of emotions given by process realization on the site of human and cobot. A lot of authors published interesting facts according organization of combined assembly processes oriented on achieving flexibility by defined production volume and product tolerances [1]. Some interesting research studies were made according to the multidimensional computing integration by production workplaces. Implementation of Industry 4.0 concept changed radically the requirements on the level of digital literacy, skills and cooperation possibilities by human and cobot. Used qualification matrix and other layout shop floor design tools were transferred in autonomously working digital managed systems. Ethical consideration of this change is now the design drivers with the limit of design space. We can see in theoretical studies, that the interconnected cyber-physical systems have integrated the new complex software and knowledge systems, oriented on the intelligent human and cobot automation, connected with digitized performance monitoring [2]. Many enterprises are working toward the goal of work optimization, in which human and cobot work together and automation is achieved through the development of intelligent machines. These intelligent self-learning systems are driven by cloud computing, breakthrough in sensor technology and the algorithms that hammers the power of bit data. Progress in robotics and artificial intelligence will change the workplace and employment law.

Important part of research contributions in this area is oriented on the human experiences and cobot support of production operations by workplace [3]. Autonomous systems have the potential save the important past process experiences, next to use these for process learning and virtual reality setting [4]. The process of workplace learning will go in new age, where the human and cobot have the same responsibilities as the pillars of stabile production process [5]. The human vulnerability can be measured through the feeling of safety by workplace and by future expectations. Main goal there is the replacement of routine human tasks through cobots and intelligent machines [6]. In recent years, we concern out attention in this area on the ethical production systems, while the production process is not only a technical matter, but collaborative human-cobot system [7]. Under systems components we manage in our digitized time the unique properties and abilities added to human and cobot, these are responsible for decision making in real time and by defined workplace [8].

Basic definition of ethics is in production process primarily oriented on the habit, which integrates the philosophy of wisdom and satisfaction with realized job output. Transformation this definition on the stabile digitized process conditions means to reflect the following parameters of human-cobot workplace:

- workplace layout—core content of value added by workplace
- production job description—knowledge of task realized by jobs
- human-cobot production operation standard—job self-realization by flexible defined production standard
- workplace productivity—“human-cobot KPI’s” competency and responsibility

An analysis of the stabile digitized workplace lead our attention on the managers, legitimate activities and technical qualification of human and cobots [9, 10]. We can call it as “biodesign” [11], this refers to new way of living organism as essential component of stabile digitized production technology. Now the motivation and improvement efforts is given by behaviour (human, cobot, information) and recognition programs are the completely impersonal processes [12].

3. Data analysis of integrated “human-cobot” workplace

Production system oriented on the implementation on Industry 4.0 concept must fulfill following parameters:

- Information background—integrated information systems should create virtual reality managed by digital technologies in real time and place
- Communication platform—machines, technologies, processes, information and material flows should connect effective virtual information world with real technological world

- Operational management—optimal performance of production system in according to the problem solving and decision making with right integration of human
- Production management—integrated value chain oriented on the standardized decision making and autonomous operability of production components
- Data security of integrated systems
- A high degree of reliability and stability of realized production processes without integration barriers
- Production systems IT maintenance
- In according to the proposal of process stabilization model we identify following testing hypothesis:

3.1 Industry 4.0 concept bring the expected cumulative benefits from digitization in the form of lower costs and increased revenues

Our survey was realized in 120 industrial companies in Czech and Slovak Republic (40% automotive, 30% mechanical, 20% external supply for automotive, 10% other industry) (**Figure 1**).

Results verified mentioned hypothesis and showed, that positive cost regulation and revenues management lies in in optimal combination of advanced connectivity (14%), better "internet of things" services (26%), computer integrated processes (35%), cloud computing and advanced automation (15%) and user-friendly process standardization (15%). Right combination of all mentioned parameters secures and bring the potential of lower cost and increased revenues.

3.2 Industrial company has identified the core processes for implementation of Industry 4.0 concept

Verification of given hypothesis has brought a clear signal to the issue of identification of key business processes. The hypothesis was not confirmed, because all companies have identified key processes mostly in relation to the production process (horizontal core processes), but not in according to the implementation of Industry 4.0 concept as supporting managerial company process (vertical processes). In according to the identification of core processes we mentioned by our survey the fact, that a lot of people in industrial companies have a problem with right understanding of the "process" definition. Traditional we are oriented on the production processes, supporting processes or we distinguish between managerial or operational processes. In the environment of Industry 4.0 concept we should make radical change in our thinking, because we are speaking about digital company—digital culture—digital processes. According to this fact we speak about new type of "process content" according to the digital enterprise environment. From our survey, we achieved secondary the following knowledge: most industrial companies have respect before traditional enterprise culture, standards infrastructure, intellectual property protection by workplace, personnel leadership and coaching. This is in direct correlation with traditional model of personnel security or personnel integrated management and decision making processes. By mentioned processes was the responsibility and delegated competencies by human, now in the Industry 4.0 concept there is necessary to transfer all important operational business

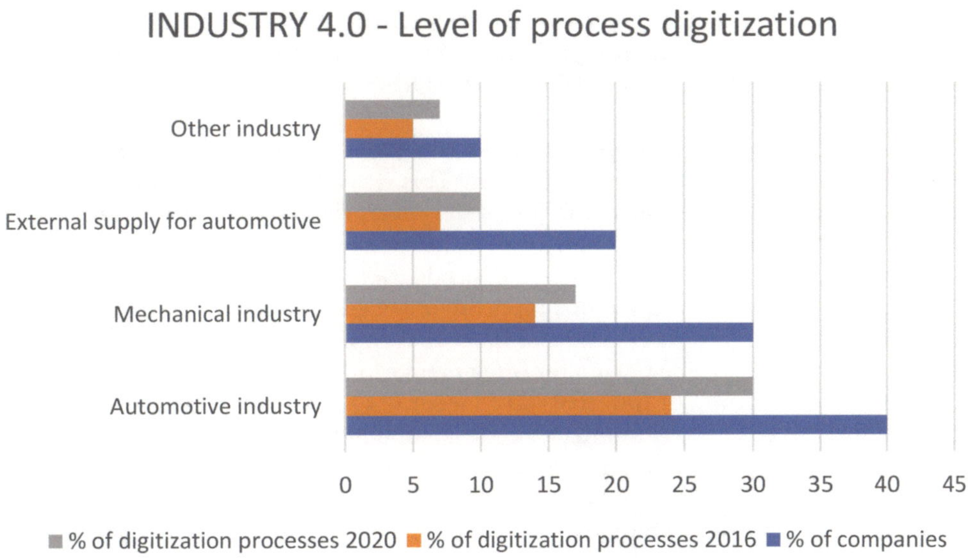

Figure 1.
Industry 4.0: level of process digitization—comparison 2016 and 2020 (source: author).

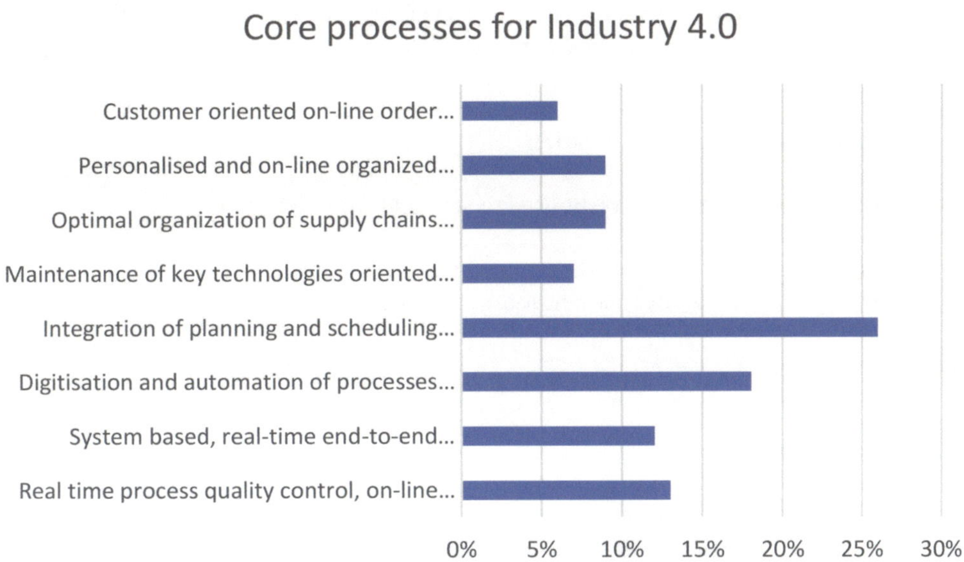

Figure 2.
Identification of core processes for Industry 4.0 implementation—process standardization and stabilization (source: author).

competencies, responsibilities connected with production processes planning, scheduling and organization to the computer technologies and digital processes (**Figure 2**).

3.3 Industrial company has a clear vision about process steps and core process stabilization pillars before the implementation of Industry 4.0 concept

A lot of companies have clear vision about content of Industry 4.0 concept, they are waiting on positive experiences from other companies, that absolved first stages of implementation process. The realized questionnaire showed the accuracy of given hypothesis with an important signal to give to the companies more knowledge and skills from successful industrial companies having experiences with first implementation steps.

Where is in enterprise competent to identify the vision and define process steps of Industry 4.0 by concept implementation? This was our principally question during realization of our research in industrial enterprises. Basic impulse for this question was the fact, that we know responsible person as a director in each enterprise for the production department (evtl. production process), but we do not know the director for implementation of supporting business processes—for example—for implementation of new IT projects. In more companies, we found during our survey product responsible person for implementation of IT project (78%), in better case for implementation of Industry 4.0 concept (14%). From point of process stabilization this is a crucial moment, because nobody from product managers cannot be responsible in industrial enterprises for combination of horizontal and vertical integrated processes. Firstly, as we can see, we should identify right person, which will be responsible complexly for the process of Industry 4.0 concept preparation and implementation (**Figure 3**).

Standardization of workplace according to the job layout, combined as "human-cobot," evokes optimal time management (possibility to cancel job when it comes to overloading or defect). It reflects the combination of integrity, security, safety, accountability, equitability and altruism components. The single most important factor in job performance and advancement is emotional intelligence [13]. Goleman mentioned in accordance to the ability to learn on the job the following parameters:

- listening and oral communication
- adaptability and creative responses to setbacks and obstacles
- personal management, confidence, motivation to work toward goals, a sense of wanting to develop one's career and take pride in accomplishments

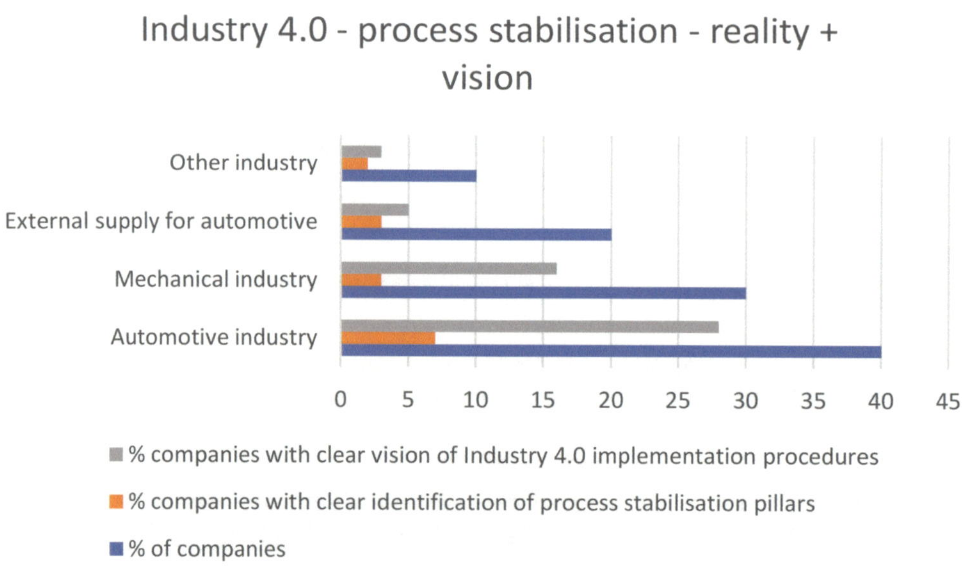

Figure 3.
Industry 4.0: reality + vision—comparison between number of industrial companies (source: author).

- group and interpersonal effectiveness, cooperativeness and teamwork, skills at negotiating disagreements
- effectiveness in the organization, wanting to make a contribution, leadership potential

Tested hypothesis:

1. **Workplace layout:** Precise job standardization and standardized layout is the most important assumption for ethical and effective "human-cobot" workplace layout (**Figure 4**)

2. **Production job description:** Knowledge of tasks realized by jobs has an impact on operative ethics and productive soft and hard cooperation between human and cobot by workplace (SMEs total number 250)

    - adequate personification of tasks realized by job and on-line organized communication and management of production processes—247 companies
    - digitization and automation of manufacturing processes—137 companies
    - flexible real-time production management by workplace—198 companies
    - on-line reception and production organization based on customer order specification—211 companies
    - continuously production flow without human obstacles or cobot conflicts by workplace—167 companies

3. **Human-cobot production operations:** People engagement into the digital organized production workplace or standardized layout design has an impact on operative ethics and collaboration with technologies (**Figure 5**)

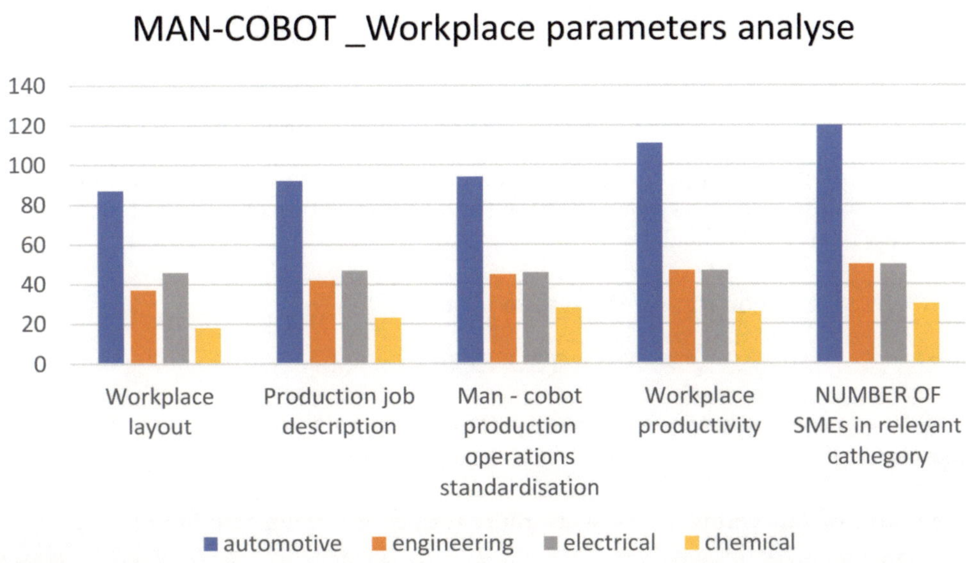

Figure 4.
Analysis of workplace layout (source: author).

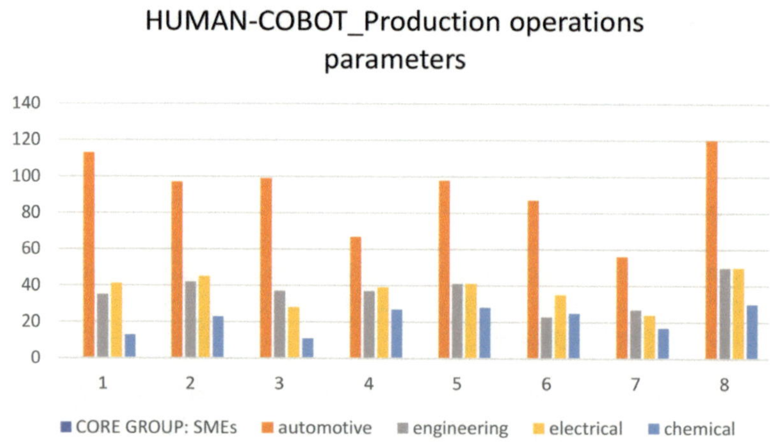

Figure 5.
Analysis of human-cobot operations (source: author): 1—comprehensive digital process map and digitized management; 2—optimal functional coordination of tasks realized by workplace; 3—defined on-line prioritization of operations and tasks; 4—flexible cooperative scheduling in real time; 7—clear standardization of ID production tasks codes; and 8—on-line digitized feedback "workplace-customer-supplier."

4. **Workplace productivity:** Virtualized knowledge and skills strong connected with the required level of mechatronics digital literacy have relevant influence on the performance KPI's connected with ethical standardized workplace "human-cobot"

Used KPI's by selected SMEs (250 companies):

- Availability of continuously production flow by "human-cobot" workplace:

$$\text{ACPF} = \text{PD–ST} - \text{PSM} \quad [\text{minute}]$$

(PD—planned time of availability, ST—service time, PSM—preventive service maintenance)

- Throughput time of e-process operation

$$\text{TePO} = \text{IITS} + \text{DeOP} + \text{WTSR} + \text{IINO} \quad [\text{minute}]$$

(IIS—instruction input time in the system, DeOP—duration of e-operation, WTSR—waiting time on system reaction, IINO—instruction input time for next e-process operation)

- Average reaction time of the system incident

$$\text{ARTSI} = \text{TSU–ID–IE} \quad [\text{minute}]$$

(TUS—time of system unavailability, ID—incident diagnosis, IE—incident elimination)

- Index of data completeness availability for process realization in Industry 4.0 system

$$\text{DCA} = \text{SRDO}/\text{RDI}$$

(SRDO—number of successful realized digitized operations given into systems as requirements, RDI—returned no realized digitized operations)

- Motivation factor for standardized "human-cobot" standardized ethical workplace:

$$MCSF = Xm_{proc} + Xm_{prod} + Xm_{pers}\ (MCSF = 1, 0)$$

(Xm_{proc}—process motivation level, Xm_{prod}—product motivation level, Xm_{pers}—personality motivation level)

0,5 level of X:
full completion of tasks integrated in human-cobot system, null level of changes or mutual personal/system conflicts (100% satisfaction).

0,4 level of X:
completion of tasks in limit given by time, amount of transactions given to the e-process or system (80% satisfaction, 20% of small disruptions or conflicts, may be waiting on the system answer or delivery of right information from other worker integrated in "human-cobot" system).

0,3 level of X:
average completion of tasks by given production plan (50% satisfaction, 50% registration of process conflicts—bad order specification, late order entry, prioritization of order in 1 hour, conflicts by persons or system settings).

0,2 level of X:
small level of mutual information exchange between staff integrated in "human-cobot" system (20% jobs realized by plan, 80% of conflicts or absenteeism of right setting of e-connections in information systems and communication channels by machines.

4. Model—stabilization of digitized processes

Flexible ethical and standardized production workplace is determined by functional "code of ethics." In many companies, we have led a highly productive discussion on the issue of the equality of working standards for workers and robots. As a result of the whole series of discussions, the standard structure of working standards for both categories should be maintained. The outcomes of these discussions were a catalog of prerequisites for the creation of an ethical standard of workplace (**Table 1**):

Next, we were identified the crucial pillars of ethical production workplace standard:

- knowledge of stabile daily structure of production program and available production technologies (elimination of daily changes in production program and higher flexibility during weekly structuring of production program)

- certainties that the stocks intended for the production process are actually available (material, staff, information, standards, layout, material flow, etc.)

- setting of clear productions and supporting processes for selected e-technology with regard to the allocation of responsibilities of specific staff (alternative for each shift individual responsible person with clear own e-code in information system)

Identification of all core process components integrated in Industry 4.0 concept
Availability of all required functionalities of computerized and digitized technologies
Number of available services integrated in digitized data and cloud system
Number of intelligent drive units integrated in the Industry 4.0 concept
Defined technical functionality, virtual functionality, communication capability and model process structure
Communication ability verification by each Industry 4.0 system component
Standardization of e-connections, standardization of input and output process parameters
Number of objects managed through digitized technology in one e-chain (ID-chain) and by one ID-administrator
Number of process conflicts in pre-implementation stage of Industry 4.0 concept (technical conflicts, technological conflicts, interface conflicts, data cybersecurity)
Number of total digitized processes in production
Number total digitized machines integrated in Industry 4.0 chain
Defined human responsibility for each process component integrated in the Industry 4.0 chain
Throughput time of production process before Industry 4.0 implementation
Throughput time of production process during Industry 4.0 implementation
Real time of system communication by realization of production process through Industry 4.0 system
React time on delivery of system components availability for realization of production process after input of customer order into system (specification and commitment of customer order)
Operative cost for order processing in digitized environment
Testing and validation of digitized processes in according to the flexible planning and organization of production process
Ability to re-plan the production process virtually by given instructions through ID competencies for flexible production organization
Availability of all relevant data and on-line data corrections availability in integrated process components
React time on process defect in system between process component owner (ID) and digitized workplace in production (identified by ID)
Grade of standardization of interfaces and abilities of units for and digitized regulation of flexible production system
Number of domain borders integrated in digitized environment for production system
Stability and security of defined standards, technological and technical rules, mutual process e-communication and e-management
Number of digital certificates for authentication of realized operations
Number of identities with login data for maintenance and management of operational Industry 4.0 system
Definition of system responsibilities for human—guaranty of system timeliness and usability
Number of virtual instances for recovery functions and security incidents elimination in Industry 4.0 system

Table 1.
Catalog of prerequisites integrated into ethical workplace standard (source: Author).

- readiness of actual dates on daily basis for production planning in information system (motivated man has all necessary information available in right structured information system, he does not need manually or with big time wastes looking for all necessary dates)

- adequate working conditions by workplace for seamless realization of production planning and control (availability of databases, knowledge of performance and technological parameters by e-machines, standards for e-oriented production planning and control, software enabled flexible production planning and control in real time, feedback from unavailable machine capacity in information system just in time)

- proper allocation of competencies and responsibilities by staff linked in a process planning and control network

- possibility (competency) to influence selected parameters of e-technologies by customer requirements in real working day (in cooperation with IT-engineer)

- real feedback from workplaces about realized production losses in information system (realized production amount, re-work pieces, re-typing times, cycle times, maintenance times, etc.)

- competency to stop the production process by the system failure and active corrections in process management system as a preventive action

- possibility of self-realization by planning—realization—control of production system in according to balancing the performance management system and innovations given for higher profitability of this performance management system (**Figure 6**)

As core processes were identified following:

- Integration of planning and scheduling processes

- Digitized automation of processes

- Real time process quality control, on-line monitoring of production and manufacturing dates

- System based, real-time and end-to-end structured processes and activities

- Personalized and on-line organized processes

- Customer oriented on-line order system

- Maintenance of key technologies oriented on on-line principles

- Personalized and on-line organized production planning, control and improvement (**Table 2**)

Emotionally feeling of actual reality by workplace may be connected with overworking, or managing our emotions. Emphatically human by workplace owns the ability to detect what his colleague is feeling. Important question: is this formula valid by robot too? We do not have the same filters that other people do to block our stimulation. As a consequence, we absorb into our own bodies both the positive and stressful energies around us. We are so sensitive that it is like holding something in a hand that has 50 fingers instead of five. What is the difference between having empathy and being an empath? Having empathy means our heart goes out to

Specification of production order	Setting up and digitizing of identification codes for product parts and relevant types of production operations Creating digital bindings of the "product—manufacturing process" Providing comprehensive digital inputs and outputs
Production portfolio management by production cell	Identification, setting up and standardization of the collaborative platform for the digitally controlled manufacturing process Flexible simulation of the production process Database of availability and performance parameters of production technologies
Production planning and scheduling	Setting up and standardizing the platform and procedures for digital decision making Personalization / Setting up the e-carrier of digital links Install and manage cloud storage of production data
Flexible job organization	Multi-level, digitally-organized production scheduling in real time Organization and administration of data analytics in real time Creating online error identification and fault elimination
Production process realization	Online Workplace Performance Monitoring Scheme Optimizing material and information flows Digital management of production processes
Customer satisfaction	Setting up a digital protocol for testing the quality of the manufactured product, the production process being carried out Creation of digitally controlled process improvement systems

Table 2.
Production standard core content (source: Author).

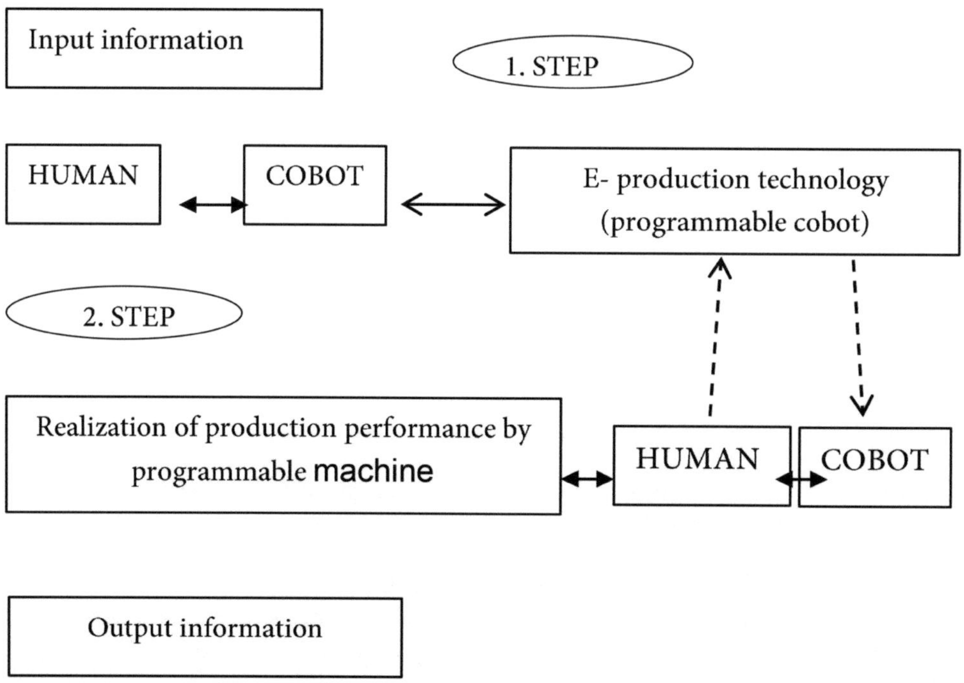

Figure 6.
Schema of mutual cooperation "human-cobot" (source: author).

another person in joy or pain. But, for empaths, it goes much farther. We actually feel others' emotions, energy, and physical symptoms in our own bodies, without the usual defenses that most people have [14].

We are supersensitive to their tone of voice and body movements. We can hear what they do not say in words but communicate nonverbally and through silence. My partner robot should understand the social situations by workplace and bring the energy to understanding of human sensitivities. Empaths are usually more successful in jobs that allow them to work in calm, well-maintained environments within a small team. Empathy goes hand in hand with emotionally intelligence. It is a complex set of skills that, when combined together, set the man up for correct relationships with other and together with yourself. And this is the first step to the self-improvement and self-motivation.

General types of empaths:

- physical empaths—human can identify physical symptoms of other people and tends to absorb them into own body
- emotional empaths—human pick up other people's emotions and can become a sponge for their feelings
- intuitive empaths—human experiences extraordinary perceptions (telepathy, intuition)

Standardly we can speak about normal empath's abilities, when the processes flow is realized continuously, without obstacles or job conflicts. If the people have not an idea, how they can collaborate or communicate in real time with cobots, they become psychologically unstable. Here it is necessary to give attention to the human sensitivity feeling moment—we should learn, how to manage standardly the human empathy, which is responsible for actual performance by workplace. In production standard is than important, to obtain real feedback to the actual experiences from loaded workflow. Man has the possibility to solve production performance defects in real time, he has also an empath to cope with the "just in moment" situation by workplace. Radically so grows his ability to react adequate according to the performance value and quality. Human is by workplace able to understand what the robot are going through in most production situations. Human empathy depends on the fact, if the automatized robot's workflow has the required working sequence as planned:

- parameters of standard construction: mind (mental state) and body (physical state)
- we distinguish next between high sensitive (high emphatic ability) and no sensitive people (low empathic ability)
- important parameter is the "depth of processing ability," which initializes by human the creative potential or ability to think in more dimensions about problem solution, high sensitive people have so good influence on the improvement processes
- big pressure on the performance or higher production value can leads to the overstimulation, which is in strong conflict with high sensitive human by workplace, is not productive and with strong influence on the surroundings **(Table 3)**

SMEs expectations—ethical aspects regarding human-cobot interaction (Analysis of 120 automotive companies CZ)					
Ethical concern	Ethical aspects	USER / OPERATOR ethical standards	MAINTAINER ethical standards	ethical behaviour of the robot itself	human behaviour (SME have ethical standards)
1st level (technical	Integrity	85—full 12—particular 23—next year	98—full 22—particular	76 SMEs accept full	46 SMEs
	Safety	113—full 7—particular	95—full 25—particular	64 SMEs accept full	92 SMEs
	Security	87—full 33—particular	89—full 12—particular 19—future	112 SMEs accept full	75 SMEs
2st level	Altruism	23—full 4—particular 93—next year	56—full 36—particular 84—future	118 SMEs accept full	24 SMEs
	Accountability	85—full 12—particular 23—next year	98—full 22—particular	76 SMEs accept full	46 SMEs
	Equitability	23—full 4—particular 93—next year	56—full 36—particular 84—future	118 SMEs accept full	24 SMEs
	TOTAL ETHICAL OPINION	78 SMEs prefer organized ethical behaviour	112 SMEs will have ethical cooperation of maintainer with cobot and operator		
Preferred attributes for "code of ethics by "human-cobot workplace"		117 SMEs—knowledge and skills of human workplace 95 SMEs—competence and prestige of human by workplace (equal opportunity for human and cobot) 99 SMEs—preparedness to realize performance 120 SMEs—association with competent persons in company (responsible for ethical workplace organization and user-friendly standardization)			

Table 3.
Analysis of SMEs expectations—ethical aspects regarding human-cobot interaction (source: author).

5. Conclusion

We collected qualitative data through semi-structured interviews with decision makers (production manager, CEOs) that are responsible are the acquisition and implementation of the cobots, workers that work side-by-side with the cobots, and through observations and demonstrations of the work system. Key requirements to the ethical standardization of "human-cobot" workplace was identified. Further research will be oriented on the next selected parameters of ethical standardized "human-cobot" workplace as automation, digitization, efficiency, intelligence with core attention given to the boundaries of digitized workplace, information by workplace, human ethical isolation (speaking with cobots), autonomy of workplace and type of knowledge used by human.

Work ethic is correlated with the intensity of firms' incentives. In societies with the wide dissemination of work ethics, firms will want effort to be complementary (as in modern production process), while the opposite is true when the dissemination of work ethic is narrow [15]. In connection to the results, presented in this paper we achieved interesting new knowledge in the area of human-cobot channels communication, identification of self-optimizing potentials by human, motion planning and job coordination as by safety, maintenance and ethical workplace standard diagnose [16–25].

Author details

Felicita Chromjakova
Department of Industrial Engineering and Information Systems, Faculty of Management and Economics, Tomas Bata University in Zlin, Czech Republic

*Address all correspondence to: chromjakova@utb.cz

References

[1] Salmi T, Marstio I, Malm T, Laine E. Man-Robot Cooperation–New Technologies and New Solutions. Micro-Assembly Technologies and Applications. IPAS 2008, IFIP–International Federation for Information Processing. Vol. 260. Boston, MA: Springer; 2008. ISBN 978-0-387-77402-2

[2] Fletcher S, Webb P. Industrial Robot Ethics: The Challenges of closer Human Collaboration in Future Manufacturing Systems. World with Robots, Intelligent Systems Control and Automation Science and Engineering. Vol. 84. 2017. pp. 159-169. ISBN 978-3-319-46667-5

[3] Wildhaber I. Artificial Intelligence and Robotics, the Workplace, and Workplace-Law. Research Handbook on the Law of Artificial Intelligence. Edward Elgar Publishing; 2018. ISBN 9781786439048

[4] Ramirez E, LaBarge S. Real moral problems in the use of virtual reality. Ethics and Information Technology. 2018;**20**(4):249-263. ISSN 1388-1957

[5] Dornberger R. Business Information Systems and Technology 4.0 – New Trends in the Age of Digital Change. Springer Verlag. ISBN 978-3-319-74322-6; 2018

[6] Espada A, Rosa C. My co-worker is a machine: A transgressive essay. In: Proceedings of the 5th Design Doctoral Conference, DDC'18, Transagression. IADE–Universidade Europeia; 2018

[7] Doyle S, Senske N. Digital provenance and material metadata: Attribution and co-authorship in the age of artificial intelligence. International Journal of Architectural Computing. 2018;**16**(4):271-280. ISSN 1478-0771

[8] Steusloff H. Humans are back in the loop! Would production process related ethics support the design, operating and standardisation of safe, secure and efficient human-machine-collaboration? In: 4th International Conference on Future Internet of Things and Cloud Workshops. 2016. pp. 348-350. ISBN 978-1-5090-3946-3

[9] Brodeur D. Work ethics and CQI. Hospital and Health Services Administration. 1995;**40**(1):111-123. ISSN 8750-3735

[10] Burns T, Stalker GM. The Management of Innovation. Kluwer Academic Publishers; 1992. ISSN 1573-0697

[11] Vettier L. Biodesign, how to think production with the living? Philosophical Readings. 2019;**11**(1): 26-32. ISSN 2036-4989

[12] Didomenico P. Business Management Daily: Low-Cost Employee Incentives, Recognition Programs and Rewards. Centerpiece, Human Resources, Performance Reviews; 2019

[13] Goleman D. Working with Emotional Intelligence. Bantam Dell Publishing; 2000. ISBN 978-0-553-37858-0

[14] Orloff J. The Empath's Survival Guide: Life Strategies for Sensitive People. Hardcover Publishing; 2017. ISBN 978-1-68364-211-4

[15] Forquesato P. Social norms of work ethic and incentives in organisations. Journal of Economic Behavior and Organization. 2016;**128**: 231-250

[16] Bendel O. Co-robots from an ethical perspective. Business information systems and technology 4.0. studies in systems, Decision and ControlSpringer Verlag. ISBN 978-3-319-74322-6

[17] Bird SJ, Spier R. Welcome to science and engineering ethics. Science and Engineering Ethics. 1995:2-4

[18] Brennan C. Digital humanities, digital methods, digital history, and digital outputs: History writing and the digital revolution. History Compass. 2018;**16**(10). ISSN 1478-0542

[19] Jocelyn S, Vurlet-Vieney, Giraud L. Experience Feedback on Implementing and Using Human-Robot Collaboration in the Workplace. Proceedings of the Human Factors and Ergonomics Society Annual Meeting. Vol. 62. 2017. ISSN 15419312

[20] Kopacek P. Development trends in cost oriented production automation. IFAC-Paper. 2018;**51**(30):39-43

[21] Li J, Tan X, Li JC. Research on dynamic facility layout problem of manufacturing unit considering human. Mathematical Problems in Engineering. 2018. ISSN 1563-5147

[22] Lin P, Abney K, Bekey GA. Robot Ethics–The Ethical and Social Implications of Robotics. Cambridge: MIT Press; 2012. ISBN 978-0-262-01666-7

[23] Sharma S, Bawa S, Lomash H. Approaches in cultural computing: A survey and interference from social computing with dynamics of mind. Wireless Personal Communication. 2018;**103**(4):2693-2713. ISSN 1572-834X

[24] Trentesaux D, Rault R. Designing Ethical Cyber-Physical Industrial Systems. Presented at the IFAC World Congress, Toulouse, France; 2017

[25] Yoo MJ, Glardon R. Manufacturing Operations Management. Europe: World Scientific Publishing; 2018. ISBN 978-1-78634-533-2

Chapter 7

Distributed Analytics Framework for Integrating Brownfield Systems to Establish Intelligent Manufacturing Architecture

Vigneashwara Pandiyan and Wahyu Caesarendra

Abstract

Intelligent manufacturing otherwise called as smart manufacturing concentrates upon optimising production and processes by making full use of data available. It is regarded as a new manufacturing model where the entire product life cycle can be simplified using various smart sensors, data-driven decision-making models, visualisation, intelligent devices, and data analytics. In the Industry 4.0 era, Industrial Internet of Things (IIoT) architecture platform is required to streamline and secure data transfer between machines, factories, etc. When certain manufacturing industry is equipped with this platform, an intelligent manufacturing model can be achieved. In today's factories, most machines are brownfield systems and are not connected to any IoT platforms. Thus they cannot provide data or visibility into their performance, health, and optimal maintenance schedules, which would have improved their operational value. This paper attempts to bridge this gap by demonstrating how brownfield equipment can be IIoT enabled and how data analytics can be performed at the edge as well as cloud using two simple use cases involving industrial robot on the abrasive finishing process. The focus of this paper is on how a scalable data analytics architecture can be built for brownfield machines at the edge as well as the cloud.

Keywords: deep learning, digital mechanical system, industry 4.0, machine learning, smart manufacturing

1. Introduction

The Industrial Internet of Things (IIoT) is set to change the face of manufacturing. Many industries are already gaining from digitalisation journey and Internet of Things (IoT), and industrial manufacturing is leading the way [1–3]. The IIoT is considered to be a modern manufacturing concept under Industry 4.0 and has been evolving rapidly based on the industrial requirement. A typical IIoT platform should consist of cutting-edge information technology (IT) infrastructure for data acquisition and sharing [4]. The features of an intelligent manufacturing includes real-time data collection and sharing among various manufacturing resources such as machines, subsystems, operators, and materials [5].

The ability to "sensorize" all and extract data to offer insights and forecasts is a huge benefit for manufacturing [6]. The Internet of Things and data remain inherently linked together. If a system goes down, for example, connected sensors can automatically localise the issue. This process is time-intensive when humans do it manually. Apart from sensorization, ubiquitous connectivity is the crucial pillar of IoT that assures to deliver the value by connecting numerous devices/assets that generate useful data. Acquiring accurate and reliable data from machines and their components is the first step in developing an IIoT architecture. Sensors might directly measure the data, which can also be obtained from controllers. Data might also be acquired from enterprise manufacturing applications such as enterprise resource planner (ERP), manufacturing execution system (MES), supply chain management system (SCMS), etc. [7]. As far as data, two important factors have to be considered. Firstly, selecting proper sensors (type and specification) for the specific objective is critical. Secondly, a seamless and tether-free method to manage the acquisition and transfer data considering various data types to the central server is required.

The data generated from IoT devices turn out to be of value only if it gets subjected to analysis, which brings data-driven analytics into the architecture. Data analytics (DA) is defined as a process, which is used to examine big and small data sets with different data properties to extract meaningful conclusions and actionable insights. These conclusions of data analysis are usually in the form of trends, patterns, and statistics that aid in effective decision-making processes. Data analysis requires support such as hardware resources such as GPU and servers for computing. The frequency of using the results of analytics may vary from real-time problems such as anomaly detection and tool wear [8, 9] to long intervals, such as predicting remaining useful life [10]. Depending on the requirement, analytics can be performed locally, i.e. edge or remotely, i.e. the cloud.

According to the aforementioned current problem statement, the IIoT-based cyber-physical system for manufacturing can fill this gap. One typical architecture of the IIoT-based cyber-physical system for manufacturing is shown in **Figure 1**. The

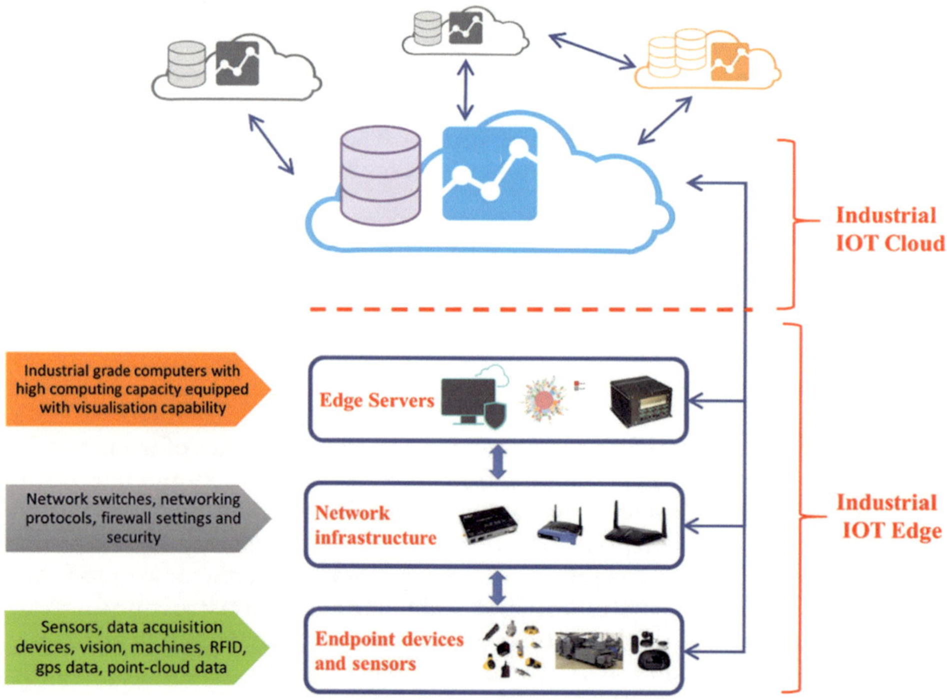

Figure 1.
Architecture of IIoT-based cyber-physical system for manufacturing [11].

first IoT layer captures data from IoT devices such as geographic coordinates, RFIDs, sensor signals, or other unstructured data. With the help of advanced computer networks, such as cyber (or software), resources and tools can be integrated with the manufacturing equipment. Connectivity is what enables the three pillar devices, data, and analytics to work. Devices must be interconnected to transfer data. Data cannot be sent or received without reliable high-bandwidth connectivity that supports real-time data flow from the many devices living on a network. Without connectivity, there would be no data to analyse and no analytics available to optimise systems and create efficiencies. IoT deployments would lack scalability. A standardised communication protocol is always required for data exchange. Many IIoT protocols can be used for networking machines, such as the Message Queuing Telemetry Transport (MQTT) protocol and the Constrained Application Protocol (CoAP). However, connectivity also exposes industrial devices to security attacks which not only disrupt entire systems but can also pose safety risks. An IIoT platform must provide security to minimise risks and keep operations protected from physical breaches and cyberattacks, by monitoring the behaviour of all data sources and alerting operators when anomalies are detected in the manufacturing environment. Securing end-to-end IIoT systems is critical in order to avoid unwanted financial and safety consequences.

The data are then stored and managed in the edge layer by IIoT platform. Like said before, it is the processing of the data acquired and stored that makes the factory realise the importance of digitalization. Data processing ranges from simple visualisation in the dashboard screen to complex cutting-edge data-driven algorithm output. High-end server or industrial PCs are generally used in this layer. Because of the cloud's ability to house large amounts of data, they are a key pillar in IIoT architecture. Today, the potential of using cloud technologies for advanced manufacturing is very high. Cloud computing can be viewed as a model for enabling ubiquitous, convenient, on-demand network access to a shared pool of configurable computing resources including networks, storage, services, and servers. Cloud computing, as well as IoT, works towards increasing the efficiency of everyday tasks, and both have a complementary relationship [12]. On the one hand, IoT generates lots of data, while on the other hand, cloud computing paves the way for this data to travel [13]. Many cloud providers take advantage of this to provide a pay-as-you-use model where customers pay for the specific resources used.

In today's factories, most of machines/systems are not connected (brownfield) and thus cannot provide data or visibility. Multinetwork environments that include ageing machines of different types and software do not speak on common connectivity language. These two pose significant challenges for many IIoT solutions and hinder the implementation of data analytics. A thorough assessment, planning, scoping, and later execution are required for implementing and performing analytics for such machines to be a part of IIoT architecture, which will be demonstrated in this research work with two use case studies.

The paper is organised as follows. A brief introduction to IIoT architecture is described in Section 1, followed by a brief overview on online quality measurement for manufacturing industries in Section 2. The experimental setup and IIoT architectural design for data analytics are presented in Sections 3 and 4. The results are described in Section 5, followed by conclusions in Section 6.

2. Output quality measurement in manufacturing process

In recent years, the development and optimization of advanced manufacturing processes have been continuously pushed to fulfil the higher demand for performance specification of the components produced. Due to a need for more consistent products

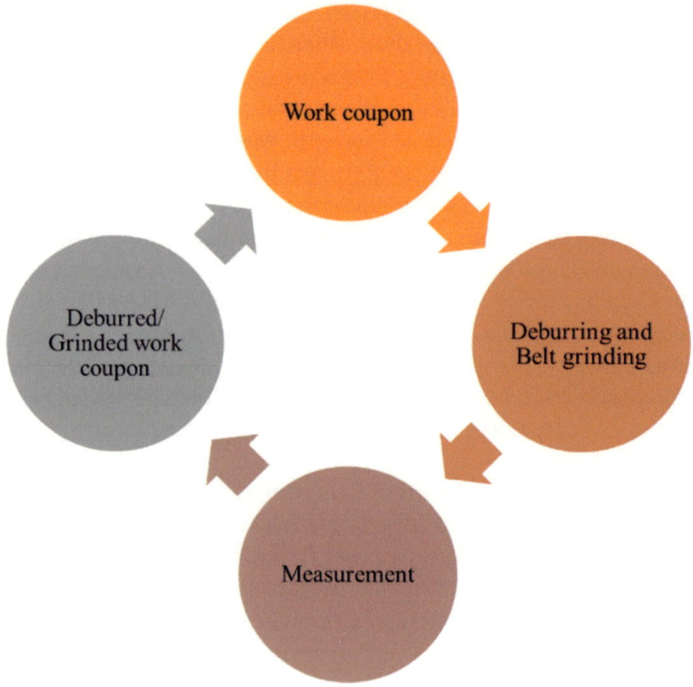

Figure 2.
Deburring/belt grinding cycle in manufacturing.

quality, product variability and product complexity, there is an increase in the manufacturing cycle and product quality measurement time. The reduction of process quality measurement requires decision making by Artificial Intelligence (AI) algorithms for which a data transfer framework for connecting between sensors, robots and devices is required. In other words, the human intervention in manufacturing line will be decreased with the development of connected intelligent manufacturing floors allowing them to concentrate more over the processes optimization, newer product designs and even maintenance. By applying an intelligent architecture to take more control over the manufacturing processes, human inconsistency and measurement time can be reduced and at the same time production capacity can be increased. These architectures will also reduce the scrap materials and enable a cleaner environment. In this paper, two case studies of the evolution of manufacturing processes from the conventional manufacturing line to smart IoT- enabled manufacturing line is presented. The processes used in our case study are deburring and belt grinding.

At present, the manufacturing industries, especially aerospace, are equipped with the industrial robots to perform manufacturing processes such as deburring and belt grinding for surface finishing. However, the output quality monitoring of surface finishing processes such as deburring and belt grinding is accomplished through conventional manual measurement. In high-volume productions, a manual measurement can lead to inconsistency from operator to operator and result in variations in product quality measurement such as thickness and surface roughness. **Figure 2** describes the deburring process cycle which involves conventional output quality measurement. As shown in **Figure 2**, which starts from the work coupon, the cycle only pauses at output quality measurement once the deburring or belt grinding process is executed. During the pause, the robot controller retracts the industrial robot to its home position. After which, an operator starts the measurement process. After the measurement is done, the cycle continues if the surface finish quality measured or required material removal is not achieved.

3. Introduction to cloud computing for smart manufacturing

In cloud computing, there are three categories of services to select from, namely, the Infrastructures a Service (IaaS), Platform as a Service (PaaS), and Software as a Service (SaaS). The sequence of these three in terms of development foundation can be described in **Figure 3**. PaaS is one layer above IaaS as platform is built on infrastructure and SaaS is one layer above PaaS as software is built on platform. The IaaS provides infrastructure consisting of computer resources and servers with network connectivity and cloud capability. Users have direct access to their servers and storage usually via application programming interface (API). The PaaS is built on a physical server infrastructure and provides a platform where users can build their software on. Meanwhile, SaaS provides ready-to-use software for users that can be accessed anywhere remotely as long as there is Internet connection. The SaaS, for example, are Google Apps, Microsoft Office 365, and Adobe Creative Cloud. Before proceeding further, the category of service has to be selected according to the requirement needed.

In building the Internet of Things (IoT) application, Software as a Service is not a relevant choice to select, unless if there is third-party software found to be useful for the application required. Thus, the selection is between IaaS and PaaS. **Table 1** lists the differences between IaaS and PaaS in the context of manufacturing application.

After considering these differences, a conclusion can be drawn. Over the consideration of long-term commitment, IaaS would be suitable for the scalability to integrate different process-specific applications into one main application. However, the scale of this application is an enterprise-level of application. A research work explains the study of IaaS architecture in more detail which can be used a useful reference for the IaaS level of implementation [14]. Meanwhile, for project level, PaaS is most suitable to build the application on. It is because in PaaS, there is no need to consider infrastructure development. Instead, the focus can be directed towards application development.

In addition to this, what cloud services are available in the market currently? In 2009 a performance comparison of several popular cloud services platform has been done [15]. The selection of cloud services should be adjusted to the time of implementation as cloud services features are always updated and have different advantages for different cloud services provider. The implementation of algorithm conversion to the cloud services provider is much easier to be done now as research platforms such as Python and LabVIEW are currently available. In addition to this difference, it is important to understand the architectural difference between the categories as explained in **Figure 4**. Hence, as PaaS is concerned, the development required to bring deburring into Industry 4.0 is on the applications and data.

The data flow of deburring and belt grinding process that includes the cloud services can be explained in **Figure 5**. In **Figure 5**, Storage Cloud Service (SCS) is

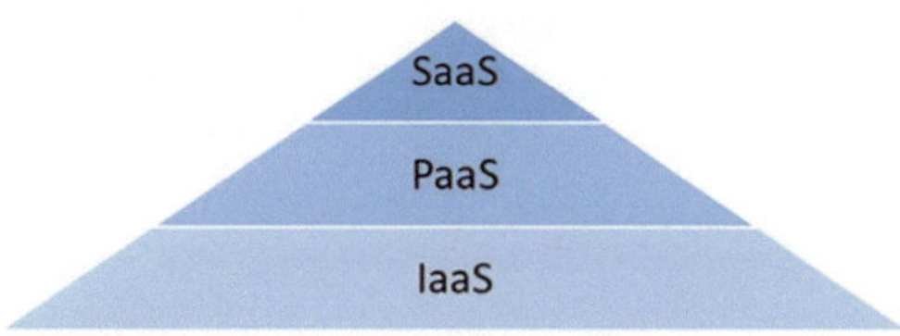

Figure 3.
Cloud services hierarchy.

Criteria	IaaS	PaaS
Developer main focus	Storage, networking, computing	Building application
Platform flexibility to build software	Flexible as developer builds the software on a platform that is on a common infrastructure	Non-flexible as developer builds the software on a platform specific to the provider
Learning curve	Higher learning curve as developer has to build both the platform and the software	Lower learning curve as developer only has to focus on the software
Scalability	Allow more flexibility	Has certain limit of scalability
Intercommunication between software	Allow software of different platforms to communicate easily on the same infrastructures	Software of different platforms must communicate to each other from different infrastructures

Table 1.
Difference between IaaS and PaaS.

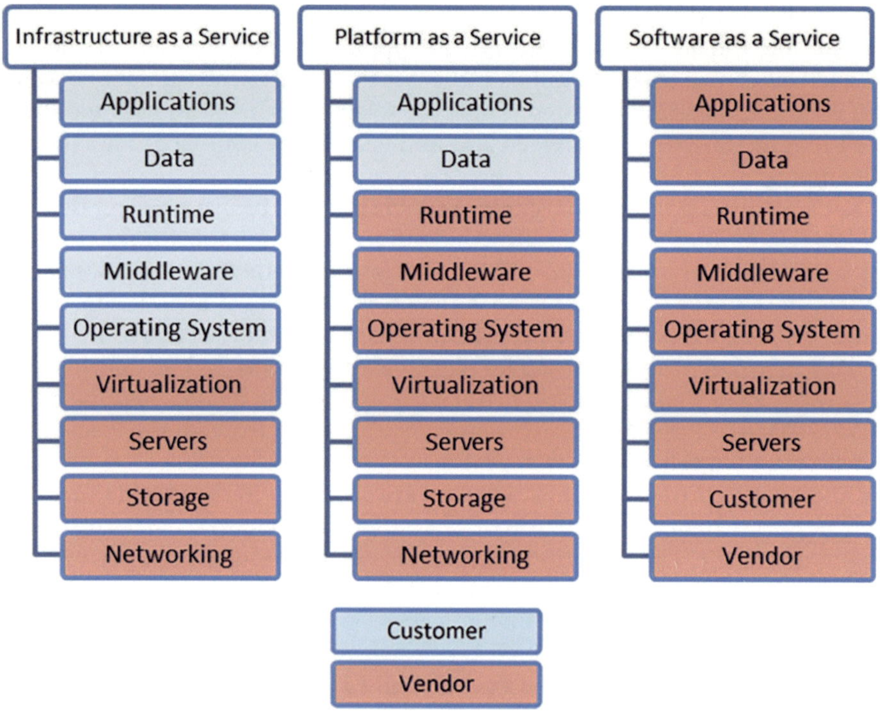

Figure 4.
Architecture difference in cloud services.

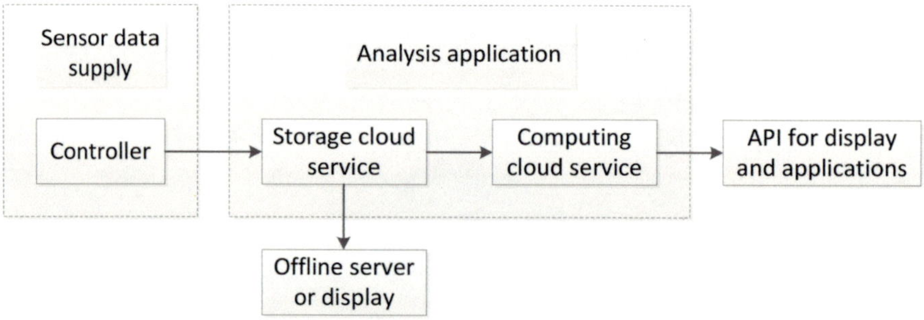

Figure 5.
Deburring and belt grinding data flow conversion from offline to online platform.

Differences	Offline data analysis	Online data analysis
Memory storage and compute capacity	Machine learning performance limited by the built-in storage capacity and memory capacity in the controller	Machine learning performance limited only by the storage and compute quota purchased under the predefined budget
Scalability	Hardware and software scale-up requirement is time-consuming	Only software scale-up requirement is time-consuming
Data resource usage and compatibility	Local data analysis using predefined development environment (e.g. LabVIEW)	Cloud data analysis using more development environment options (e.g. TensorFlow, Kaffe2, Keras)

Table 2.
Implementation difference between offline and online data analysis.

the storage service to store the sensor data and the result in the cloud. Meanwhile, the computing cloud service (CCS) is the computing capacity to process the data. Thus, the machine learning is adapted to the cloud through CCS. This architecture provides API output for display or other applications when required. In addition, the data in SCS is also tapped into offline server or display when required.

The difference in implementation between offline and online data analysis through the same machine learning model is described in **Table 2**, assuming the use of cloud services by third party.

4. Experimental setup (case study)

The robot-assisted manufacturing processes are gaining popularity in industries that move towards automation. Integration of the processes on the robot helps in easing the axis of motion and also ensures that the force applied and repeatability is maintained [16, 17]. Abrasive belt grinding and deburring processes are typically combined with an industrial robot in the manufacturing industry for achieving the desired surface finish and tolerance. In this paper, we will be demonstrating how these two brownfield manufacturing systems can be IoT enabled, and analytics can be performed at two different levels, i.e. cloud and edge. The process data from deburring will be used for edge analytics, and data from belt grinding will be used to perform machine vision analytics at the cloud.

4.1 Deburring

The fundamental of Industry 4.0 is the data communication. In deburring, the main data communication path happens between (1) the sensors to the DAQ (National instrument – Compact RIO) controller, (2) the DAQ controller to the robot controller, (3) robot controller to the ABB robot, and (4) DAQ controller to the server and cloud. The data communication is done through digital and analog input/output with the maximum analog transfer rate at 4 MS/s per module four 16-bit (64 kb) analog input. This translates to maximum of 256 Mb/s data transfer rate per module if the module is fully utilised to its capacity.

Figure 6 explains the data flow in the deburring process. The data transfer occurs locally from the sensors and DAQ controller initially. DAQ controller preprocesses the sensor data and arrives at a decision based on the trained AI model. The DAQ controller feeds the decision output back to the robot controller via its digital-analogue I/O. In the branch of this local control loop, the preprocessed data is sent to the cloud in a suitable protocol with data encryption.

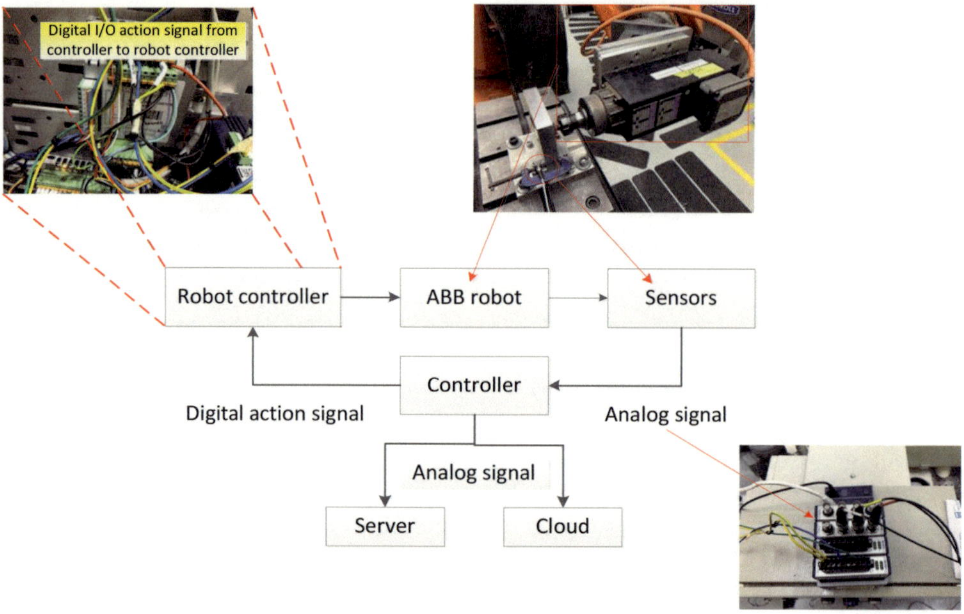

Figure 6.
Data path loop and DAQ to robot communication in deburring.

The implementation of data analysis to the cloud requires conversion of the machine learning algorithm to the cloud. Before moving into this, some fundamentals of the cloud services for Industry 4.0 implementation are explained to understand which platform is suitable for certain application.

4.2 Belt grinding

The current industrial practice of removing weld seam in manufacturing industry involves skilled operator using a belt sander. The component, when removed from the production for such operation, poses delay to the production cycle; fixturing, clamping, and unclamping cause a loss to the production volume. As an alternative, the belt grinder is integrated with the robot (in our case ABB6640). The tool path is programmed in the robot controller in such a way that weld seam and surrounding areas are blended. Machine vision-based solution is opted to analyse the state of the weld seam. If the weld seam is yet to be removed, the controller is triggered via flag setting through input/output pin. If the weld is removed, the controller is not triggered, and the completion message is sent.

After completion of each pass, the IP camera is triggered to capture the image of the component. The edge Pc that triggers image capturing knows the state of tool path as it is the constant communication with the robot controller via TCP/IP. The image acquired is then passed through a secured network on MQTT protocol to a MATLAB instance running on a virtual machine inside Azure cloud services. The MATLAB instance has the deep learning model-based encoder-decoder architecture on predicting the pixel-wise state of the weld seam. The result is sent back to the edge PC, which initiates the tool path again. The cycle continues as long as the weld seam is completely blended with the surrounding surface. The deep learning algorithm is trained using four different weld seam states. The edge PC triggers the robot controller to execute the tool path unless weld seam state four is reached, i.e. where the weld seam is wholly removed. The schematic flow of data connectivity and decision-making is shown in **Figure 7**.

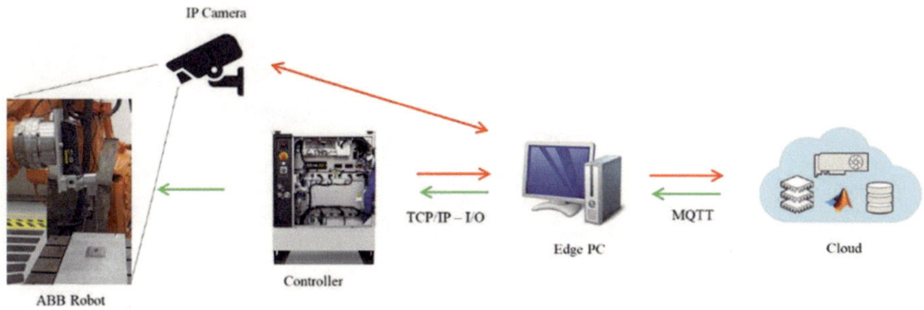

Figure 7.
Data analytics for the IIoT architecture in the cloud for weld seam prediction.

5. Results and discussion (case study)

5.1 Deburring

In order to establish data collection and analysis through machine learning to predict the deburred surface quality, several equipment have to be prepared. **Table 3** lists the systems used in this paper. Meanwhile, the system configuration for deburring is described in **Figure 8**. In **Figure 8**, it can be seen that the controller is the core equipment whose functions are to collect the sensor data, analyse, control robot action, and upload the data to the server and cloud. Thus, DAQ controller specification is crucial to determine the performance of this framework adaptation for deburring in the production line.

After understanding the data collection, the next step is to understand how to implement machine learning model into deburring process. A schematic is made to understand how the machine learning model processes the sensor data in deburring and predicts the physical features such as chamfer length and surface finish.

Figure 9 is the training phase to generate a model that correlates the input sensor data in deburring to the physical features measured on deburring work coupon as output. After the trained model satisfies the accuracy and repeatability required, the trained model is again implemented between the input and output. In the author's previous journal paper, an example is discussed for the use of Welch's estimate to compute power spectral density (PSD) to classify the number of passes and classify the vibration signal generated by the spindle based on adaptive neuro-fuzzy inference system (ANFIS) [18]. Meanwhile, in the second journal publication of this topic, the authors presented the results of fuzzy inference system (FIS) machine learning method to obtain the corresponding output as predicted surface finish quality of boss hole chamfer length and also the stage classification of deburring process [19]. A detailed result on feature extraction and machine learning

Equipment	Function
ABB robot	Robot for machining
DAQ Controller	Sensor data acquisition, analysis, and data transmission to the Internet
Sensors	Collecting the variables required from the process for analysis
Display	Displaying the sensors data, analysis process, and result

Table 3.
System for industry 4.0 implementation.

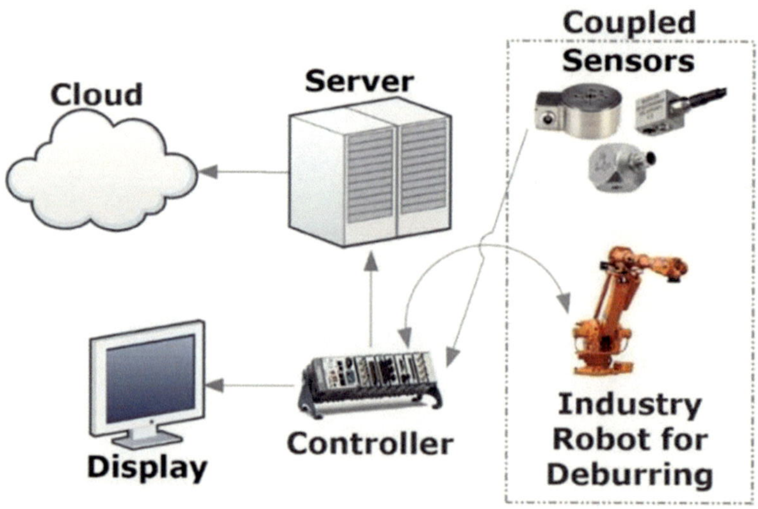

Figure 8.
Deburring equipment configuration and data path.

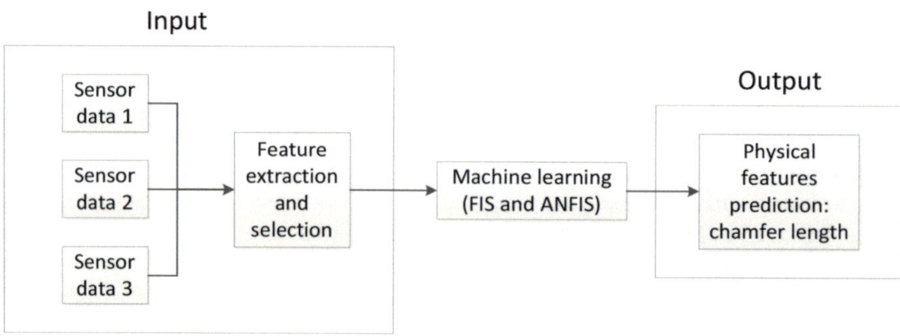

Figure 9.
Fuzzy inference system analysis model training on deburring.

method using ANFIS and FIS can be obtained from previous publications [18, 19]. A preliminary study of cloud computing to predict the output quality of deburring is presented in [19].

5.2 Belt grinding

An IP camera is introduced to capture the images of various stages of the weld seam. The IP camera is incorporated with the help of a tripod stand adjacent to the belt grinder. Surface images are captured at the end of every pass of robot arm across the weld seam (**Figure 10**). The camera system is capable of capturing and storing the surface images at a resolution of 1240 × 960 pixels. The images are subsequently labelled and sent to the Azure cloud with MATLAB Environment where the actual model training takes place. The variable grinding parameter used to remove weld seam is shown in **Table 4**.

Azure's in-built architecture incorporates best practices for creating a full MATLAB desktop experience on Azure. This includes connecting to Azure from your local desktop using Remote Desktop Protocol (RDP). It sets up a single virtual machine containing MATLAB, a private virtual network with an Internet

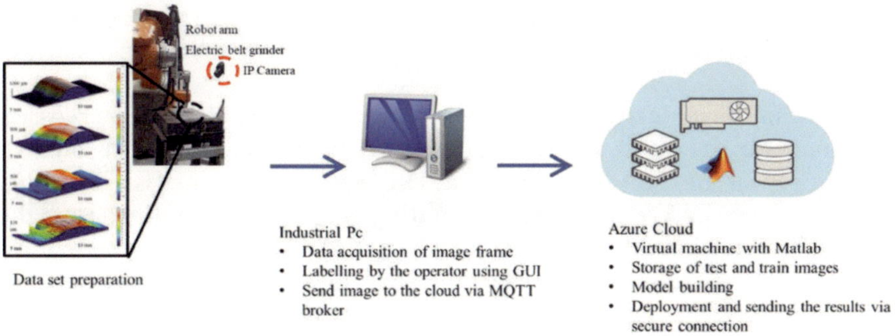

Figure 10.
Analyse data in the Azure cloud with MATLAB [20].

Parameter	Description
Belt grinding speed	5000–1100 RPM
Contact wheel diameter	10 mm, 24 mm
Hardness of contact wheel (polyurethane)	Shore A Hardness 30, 60, 90
Lubrication	Dry condition
Feed	10–40 mm
Belt finishing duration	Variable time
Operational mode	Position control

Table 4.
Parameters used in the belt grinding experimental trials.

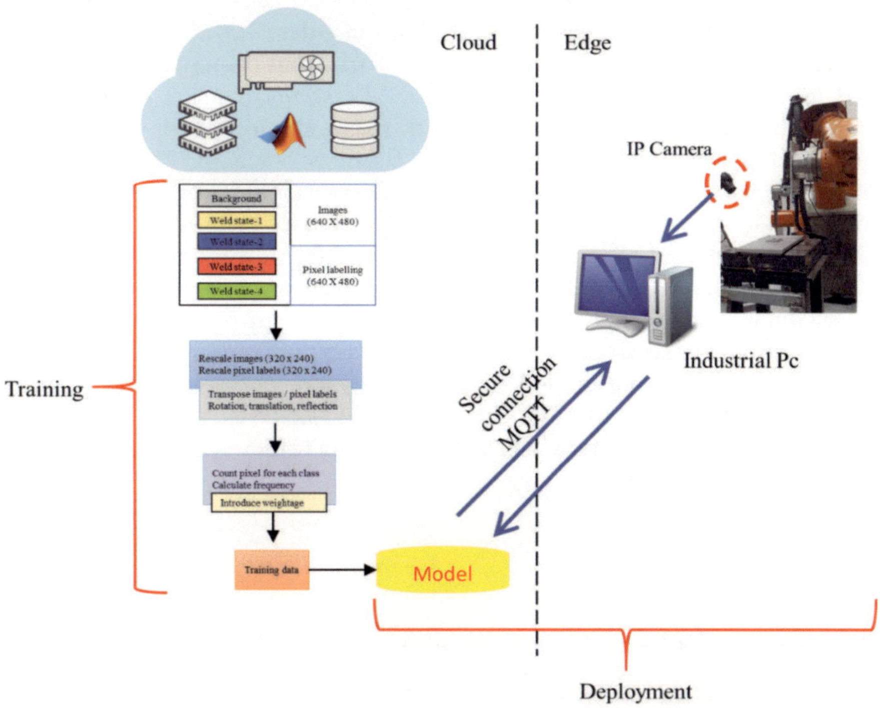

Figure 11.
General description of the proposed methodology to train and deploy the model in azure cloud.

gateway, a private subnet, and a security group that opens the appropriate ports for SSH and RDP access. For training and deploying, we will be using Azure cloud service as IaaS.

Figure 11 shows in detail the implementation of encoder-decoder-based deep learning model in the cloud. In total, 2000 images taken using the IP vision system are labelled offline and stored in the cloud. The MATLAB instance is created in the cloud, and required resources for computing are generally chosen before the creation of VM on which the MATLAB instance runs. The deep learning model is defined to identify four weld seam states. The VGG-16 network is retrained to identify the weld seam states. Pixel label layer of the default VGG-16 network is replaced by the customised label layer that would identify the weld seam state and background of an image. Distribution of pixel count for four different belt states and background is identified, and the corresponding weight is redefined on the final layer of the VGG-16 network. The weld state identification was performed in MATLAB deep learning toolbox. The augmented training image set is used for training, and it is ensured that the training accuracy increases and training loss decreases with the iteration count. The training is terminated once the parametric conditions are met. Once the model is ready for deployment, IP camera sends the image frame at an interval of the 1-second interval through secure MQTT to the cloud during the actual belt grinding trials. The script in the cloud passes the image into the developed model to make the pixel-wise classification. It is to be noted that the transfer speed of images from edge to cloud depends on the network traffic. The pixels' classified image is sent back to the edge PC for visualisation through MQTT secure network.

6. Conclusions

Two case studies related to the machine learning, deep learning, and cloud computing application towards smart manufacturing has been presented in this work. A methodology for integrating brownfield systems into IIoT framework to facilitate industry 4.0 adaptation is demonstrated. A simple data flow pipeline has been established between the edge and cloud framework via MQTT protocol. Cloud framework based on IaaS is used to deploy the deep learning model. However, it is to be noted that other protocols similar to MQTT can also be used for data transfer. The deployment of a robust analytical architectural framework cannot be just restricted to putting data processing and analysis software in place at cloud and edge. As a plant expands, the analytical framework and supporting hardware need to evolve organically. The framework should also support repeated installations and setup procedures being carried out simultaneously for brownfield equipment as well as systems with in-built IoT. Caution should also be exercised while amalgamating the analytical solution in cloud for existing legacy systems, in order to preserve their security and integrity.

Conflict of interest

The authors declare no conflict of interest.

Author details

Vigneashwara Pandiyan[1] and Wahyu Caesarendra[2*]

1 Smart Manufacturing Group, A*star Advanced Remanufacturing Technology Center, Singapore

2 Faculty of Integrated Technologies, Universiti Brunei Darussalam, Brunei Darussalam

*Address all correspondence to: wahyu.caesarendra@ubd.edu.bn

References

[1] Shariatzadeh N, Lundholm T, Lindberg L, Sivard G. Integration of digital factory with smart factory based on internet of things. Procedia CIRP. 2016;**50**:512-517

[2] Jeschke S, Brecher C, Meisen T, Özdemir D, Eschert T. Industrial internet of things and cyber manufacturing systems. In: Industrial Internet of Things. Cham, Switzerland: Springer; 2017. pp. 3-19

[3] Zhou K, Liu T, Zhou L. Industry 4.0: Towards future industrial opportunities and challenges. In: 2015 12th International Conference on Fuzzy Systems and Knowledge Discovery (FSKD), IEEE. 2015. pp. 2147-2152

[4] Chen Y. Integrated and intelligent manufacturing: Perspectives and enablers. Engineering. 2017;**3**(5):588-595

[5] Zhong RY, Xu X, Klotz E, Newman ST. Intelligent manufacturing in the context of industry 4.0: A review. Engineering. 2017;**3**(5):616-630

[6] Stojanovic L. Intelligent edge processing. In: Machine Learning for Cyber Physical Systems. Berlin, Heidelberg: Springer; 2020. pp. 35-42

[7] Domingo Galindo L. The Challenges of Logistics 4.0 for the Supply Chain Management and the Information Technology. Norway: NTNU; 2016

[8] Chen X, Limchimchol T. Monitoring grinding wheel redress-life using support vector machines. International Journal of Automation and Computing. 2006;**3**(1):56-62

[9] Pandiyan V, Caesarendra W, Tjahjowidodo T, Tan HH. In-process tool condition monitoring in compliant abrasive belt grinding process using support vector machine and genetic algorithm. Journal of Manufacturing Processes. 2018;**31**:199-213

[10] Zheng S, Ristovski K, Farahat A, Gupta C. Long short-term memory network for remaining useful life estimation. In: 2017 IEEE International Conference on Prognostics and Health Management (ICPHM), IEEE. 2017. pp. 88-95

[11] Jiang J-R. An improved cyber-physical systems architecture for industry 4.0 smart factories. Advances in Mechanical Engineering. 2018;**10**(6):1687814018784192

[12] Xu X. From cloud computing to cloud manufacturing. Robotics and Computer-Integrated Manufacturing. 2012;**28**(1):75-86

[13] Tao F, Cheng Y, Da Xu L, Zhang L, Li BH. CCIoT-CMfg: Cloud computing and internet of things-based cloud manufacturing service system. IEEE Transactions on Industrial Informatics. 2014;**10**(2):1435-1442

[14] Bhardwaj S, Jain L, Jain S. Cloud computing: A study of infrastructure as a service (IAAS). International Journal of Engineering and Information Technology. 2010;**2**:60-63

[15] Peng J, Zhang X, Lei Z, Zhang B, Zhang W, Li Q. Comparison of several cloud computing platforms. In: 2009 Second International Symposium on Information Science and Engineering. 2009. pp. 23-27

[16] Ren X, Cabaravdic M, Zhang X, Kuhlenkötter B. A local process model for simulation of robotic belt grinding. International Journal of Machine Tools and Manufacture. 2007;**47**(6):962-970

[17] Ren X, Kuhlenkötter B. Real-time simulation and visualization of robotic belt grinding processes. The

International Journal of Advanced Manufacturing Technology. 2008;**35**(11-12):1090-1099

[18] Caesarendra W, Wijaya T, Tjahjowidodo T, Pappachan BK, Wee A, Roslan MI. Adaptive neuro-fuzzy inference system for deburring stage classification and prediction for indirect quality monitoring. Applied Soft Computing. 2018;**72**:565-578

[19] Caesarendra W, Pappachan BK, Wijaya T, Lee D, Tjahjowidodo T, Then D, et al. An AWS machine learning-based indirect monitoring method for deburring in aerospace industries towards industry 4.0. Applied Sciences. 2018;**8**(11):2165

[20] MATLAB and MATLAB in the Cloud. Natick, Massachusetts, United States: The MathWorks, Inc.; 2018b